I0814148

THE FIREARM REVOLUTION

The Firearm Revolution

FROM RENAISSANCE ITALY TO THE EUROPEAN EMPIRES

Catherine Fletcher

PRINCETON UNIVERSITY PRESS
PRINCETON & OXFORD

Published by Princeton University Press
41 William Street, Princeton, New Jersey 08540
99 Banbury Road, Oxford OX2 6JX

press.princeton.edu

GPSR Authorized Representative: Easy Access System Europe - Mustamäe tee 50, 10621 Tallinn, Estonia, gpsr.requests@easproject.com

ISBN 9780691272672
ISBN (epub) 9780691289335
ISBN (PDF) 9780691272696

Library of Congress Control Number: 2025949892

British Library Cataloging-in-Publication Data is available

Editorial: Ben Tate and Josh Drake
Production Editorial: Mark Bellis
Jacket Design: Katie Osborne
Production: Danielle Amatucci
Publicity: Alyssa Sanford and Carmen Jimenez

Jacket image: (front): Universal Images Group North America LLC / Alamy. (Spine): Unknown maker, priming flask/spanner, France, c. 1570–80. Steel, gold, silver, 14.9 cm length. Gift of William H. Riggs, 1913. Metropolitan Museum of Art, 14.25.1523. Public domain.

This book has been composed in Miller

Printed in the United States of America

10 9 8 7 6 5 4 3 2 1

To Sandra Cavallo

CONTENTS

LIST OF ILLUSTRATIONS

Figures

Plates

A NOTE ON MONEY

A RANGE OF currencies circulated in sixteenth-century Italy: each city-state issued its own and during wartime exchange rates were far from stable. Inflation was also a significant issue, and many people received elements of wages in kind. Due to these complexities, I have avoided giving modern equivalents for sums of money mentioned in the book, since they raise as many questions as answers. It is, however, relevant to know that the key coins in circulation were ducats; these were gradually superseded by the scudo, worth about 6 per cent less. Venice had a silver coin known as the lira, but the system of lire, soldi and denari (pounds, shillings and pence) was used for accounting purposes more widely in Italy, even where the principal coin was the ducat. Ducats were worth about 7 or 8 lire. Pikemen were paid up to 3.5 ducats a month, but late payment was common; arquebusiers could earn twice that but had more expenses to cover. Further details may be found in Carlo Cipolla, *Money, Prices, and Civilization in the Mediterranean World: Fifth to Seventeenth Century*, and Michael Mallet and Christine Shaw, *The Italian Wars 1494–1559: War, State and Society in Early Modern Europe*, 209–11.

THE FIREARM REVOLUTION

Introduction

THE GUN IS not the first object to come to mind when one thinks of Renaissance Italy. Perhaps it should be. In 1433, as Cosimo de' Medici was forced into exile from Florence, the first handguns were spotted on the peninsula. Guns were used at the 1440 Battle of Anghiari, and later in the century Leonardo da Vinci sketched a design for a wheel lock that would allow a gun to be fired without a lighted match. The Beretta company dates itself back to 1526, the year before Machiavelli's death and when Michelangelo was working on the Medici funerary chapel at the basilica of San Lorenzo in Florence.[1] Later that decade, Paolo Giovio, who combined careers as a historian and physician at the papal court, identified print and gunpowder, along with the voyage of Columbus to the Americas, as inventions that had 'made this age so renowned for its good fortune'.[2] Others followed in that assessment of gunpowder's significance to modernity, notably Girolamo Cardano (in his 1575 autobiography),[3] Jan van der Straet and Philips Galle, in the frontispiece of their *New Inventions of Modern Times*, and Francis Bacon.[4] Firearms, however, are also central to problematizing ideas of progress, representing as they did new possibilities for violence and conquest. In the eighteenth century, guns would be regarded as a more gentlemanly weapon, thanks to the possibility they presented of avoiding a direct physical fight.[5] In this sense, they were sometimes characterized as relatively civilized, a description that plays down their comparative lethality. In the sixteenth century, however, guns were still regarded as unchivalrous, unmanly, and diabolical: they still had to be 'domesticated' and made socially acceptable, and masculine identities had to be modified to accommodate this new model of the armed self.[6] That process took some time. To borrow one of the many gun-related phrases that pervade the English language, Europe shot first and asked questions later.

The continent's debates about firearms take us to the heart of an argument about how and whether early modern Europe underwent a 'civilizing process'.[7] The imagined golden age of the Renaissance, described by Baldassarre Castiglione in his 1528 *The Book of the Courtier*, is one of learned men and women eloquently debating questions of art and beauty in their castle gardens, a world in which, while arms remained the 'first profession' of the courtier, nobility came to be constituted as much through virtuous knowledge of letters as through prowess on the battlefield. Yet even while royal courts became more settled, the life of kings less characterized by direct participation in battle, and the upper classes more restrained and polite in their bodily comportment, guns became ever more common. Research on seventeenth-century Italy has shown just how fragile this process of civilizing was, and how easily even major cities could fall into patterns of violent feuding.[8] Society could be simultaneously more modern and more lethal.

The tensions around guns become apparent in the case of a cowherd named Vincenzo, who in 1552 obtained a licence from the Bologna authorities that allowed him to carry arms, including a gun for the purpose of hunting.[9] It came with the standard caveat that the weapon must not be carried at churches, festivals or in markets. These were busy environments where an accidental misfire had higher risks than out in the countryside, and where a generalized ban on arms-bearing (staff weapons such as pikes were also included) made policing more straightforward. The local guard did not have to decide whether any given individual was plotting a murder in the cathedral: they could simply exclude anyone carrying a gun. Vincenzo was taking advantage of a technology that was proliferating rapidly. Fifty years before, handguns had been a relative novelty on the Italian peninsula, but after decades of conflict they had become far more familiar: the Italian Wars (1494–1559), in which France and Spain contended for hegemony on the peninsula, saw firearms become decisive on the battlefield for the first time.

The rulers of Italy's multiple states promoted shooting in the interests of civic defence and military preparedness, but as early as 1522 the same rulers began to issue gun control decrees. These attempts at regulation, however, often failed to convince the population. The anonymous author of a proposition for international gun control observed, probably in the late 1570s, that

> if the Pope, as chief, and then the other princes do not make some prompt provision, life will be so badly corrupted in this matter that it will become even more difficult, and soon there will be no place nor state where people are safe, given that every low herdsman or shepherd you meet in the countryside today has a wheel-lock arquebus over his shoulder.[10]

In these first centuries of guns, pressure for gun control in the face of proliferation was already apparent.

In fact, wheel-lock firearms, particularly those of smaller sizes, were extensively regulated in sixteenth-century Italy. Unlike the traditional matchlock (which worked by application of a lighted match to gunpowder), they were fired by means of a self-lighting mechanism, could be concealed beneath clothing, and were thus a handy weapon for assassins and bandits. Rulers were intensely concerned by their threat to public order. Yet, at the same time they could see a case in favour of firearms. Cardinal Ferdinando de' Medici, to whom that gun control proposition may have been addressed, had in his library a manuscript known as the *Florentine Codex*. It described in some detail the advantage that guns conferred on the Spanish commander Hernando de Cortés and his troops as in 1521 they entered the Aztec capital Tenochtitlan:

> The fifth group [of the army] were those with arquebuses, the arquebusiers. . . . And when they had come to enter the great palace, the residence of the rulers, they fired them; they repeatedly fired the arquebuses. They each exploded, they each crackled, were discharged, thundered, disgorged. Smoke was spread, smoke was spread diffusely, smoke darkened, smoke massed all over the ground, spread all over the ground. By its fetid smell it stupefied one, it robbed one of one's senses.[11]

How rulers in Italy and beyond balanced their desire for civic defence and overseas conquest with the maintenance of social order—and the impact their decisions had on wider society—is the subject of this book.

A study of guns, moreover, brings new insights to our understanding of many wider social developments in sixteenth-century Europe. The Italian Wars drew in combatants from across Europe as France and Spain fought for hegemony on the peninsula and, as we will see, ideas about gun use were exchanged between men (and it was primarily men) of different nationalities. Guns were everywhere in this society. They can tell us about courtiers and other social elites, and about how the private companies that produced them interacted with the state and the older feudal aristocracy. They can tell us about perceptions of gender, rank and race. Efforts to regulate gun use in the later sixteenth century can be read in the context of post-Reformation state formation, as rulers on either side of a continent divided into Protestants and Catholics sought to prove their commitment to orderly Christian morals. In relation to European empire-building, firearms briefly offered a technological advantage: in early accounts of global encounters stories about guns underlined a narrative of European superiority. In fact, however, guns were swiftly adopted and assimilated by both Africans and Native Americans. In contrast, during conflict with the Ottoman Empire, the Catholic powers faced an opponent with a well-established firearm industry, making for somewhat, yet not entirely, different portrayals of gunpowder weapons.

Firearms, however, do not only offer a lens through which to rethink early modern European society. This is a topic with significant contemporary resonance. Sixteenth-century Italy was a society where the risk of conflict was—and was perceived to be—high. Divided into multiple and often warring states, its relative wealth and location at the crossroads of the Mediterranean meant that foreign powers were keen to secure their influence on the peninsula. To do so they sought out local allies. The modern distinction between soldier and civilian was not acknowledged in the contemporary laws of war, and non-combatants were commonly targeted; beyond the military context, banditry was an established problem.[12] It is hardly surprising that both states and individuals saw a case for arming themselves, yet at the same time they recognized the risks. Many of the arguments raised today in relation to gun control are to be found in sixteenth-century sources. These include calls for restrictions on the ownership of those weapons judged most dangerous, demands from users that they be allowed to keep guns for self-defence (but also critiques of firearms' effectiveness in defensive contexts), associations with honour and, perhaps most notably, the question of how to regulate an armed militia. It will become apparent to readers that these early modern documents not only echo debates in the United States, but also discussions in post-conflict environments and in societies and subcultures worldwide where honour remains an important justification for firearm ownership.[13] Indeed, when I first wrote on this topic, during a master's degree at the start of this millennium, the United Kingdom had recently had its own debate on gun laws. This followed the tragic massacre at Dunblane Primary School, where on the morning of 13 March 1996 sixteen children and their teacher were shot dead. I had attended high school in Stirling, the nearby county town, so these questions came rather close to home. In 1997, most handguns were banned in England, Wales and Scotland. The one part of the United Kingdom that retained more latitude for gun ownership was Northern Ireland, where in acknowledgement of its decades of conflict it is still legal, if uncommon, for handgun permits to be issued for self-defence. Yet if firearms now rarely hit the UK domestic headlines, coming back to the topic after more than two decades, I remain struck by the wider continuities between past and present.

Firearms, moreover, have a further interest for anyone concerned with the impact of new technology and new products on society. Much consideration has been given to the change wrought by printing on Europe: in the four and a half decades since Elizabeth L. Eisenstein called the printing revolution unacknowledged, it has been the subject of multiple studies.[14] For the eighteenth century, historians have explored how new commodities imported from European empires—coffee, chocolate, sugar, tea, tobacco—changed the social and cultural life of the metropole. Looking back on that literature, which I taught to first-year undergraduates twenty years ago, I now wonder why guns were not on the list. Considerations of the 'gunpowder revolution' have tended to confine it to a military sphere. Yet it should not be a surprise to find technologies originally developed for military purposes finding wider uses. We might be sceptical now about the extent to which that was desirable, but Vincenzo the cowherd had a practical use for his gun beyond its use in warfare. Once the genie was out of its bottle, there was no putting it back in.

None of this is to say that scholars have ignored the history of early modern firearms. Until recently, however, guns were typically to be found in the margins of any given field of study. Military historians considered their impact on the battlefield. Historians of violence noted their use in crime. Studies aimed at curators and connoisseurs documented makers' marks, materials and technological developments in detail.[15] Overall, however, there were few broad analyses to be found until the global survey by Kenneth Chase published more than two decades ago. Interest has grown since then, as scholars focusing on England and the German states have established the significance of firearms in domestic and civic contexts, and especially the importance of militia service in their proliferation. Lois Schwoerer has made a case for the existence of a gun culture in England, noting the rising interest in firearms at the court of Henry VIII, who was shooting with handguns by 1536–37 and arranged additional tuition for himself in 1541; B. Ann Tlusty and Jean-Dominique Delle Luche, on the other hand, have shown how firearms were assimilated into existing martial cultures in the Holy Roman Empire.[16] Sheila Nayer (for England) and Patrick Brugh (for Germany) have explored the literary responses, which are largely hostile.[17] Until recently much literature on Italian handgun production focused on the identification of gunfounders, although there is also an important history of the Beretta firm, and local histories of Valtrompia and the nearby city of Brescia provide further information.[18] A 2022 volume produced to mark the thirtieth anniversary of the Luigi Marzoli museum of arms and armour in Brescia has proven a valuable addition, while the work of Michael Mallett and John Hale on the military organization of the Venetian state (now forty years old) sets these studies in wider context, as does Hale's essay on gunpowder and the Renaissance, and more recent studies of the Italian Wars.[19] In these broader analyses, the role of arms manufacturers and brokers, who served a clear purpose for the state in the military context but also had their own interest in promoting weapons use and encouraging the expansion of firearms technology, has often been lost, but this has begun to be remedied, with recent studies exploring the arms industry in seventeenth-century Brescia; artillery production in Florence; and the gunpowder industry in Venice. To these should be added the work of David Parrott, who takes a broader European view on the question of military enterprise.[20]

New scholarship on material culture offers a further perspective through which to study handguns. There is now a rich body of literature on Renaissance Italy that draws on the peninsula's extensive archive record to examine everyday life through objects, considering for example practices of gift-giving, and the location of things in domestic and institutional space.[21] Growing numbers of scholars have turned to this approach to understand the history of war, although their studies have often focused on objects ancillary to conflict rather than weapons.[22] Notable exceptions are Victoria Bartels, whose work on masculinity in sixteenth-century Florence situates arms and armour in a wider gendered and cultural context, and Kristen B. Neuschel, who explores swords over the longue durée in France and Britain.[23] Building on this scholarship, I have paid close attention to the gun as an object, exploring its life cycle from production to decay, and considering its materials, users and meanings.

Alongside individual interactions, throughout this book I discuss guns in relation to the state, a term that needs some qualification. The concept of the state in Italy is not straightforward, uniting as it does polities of different sizes and political types, ranging from more or less broad-based republics to principalities, as well as those areas ruled as subject territories. Though commonly translated as 'state', the term *stato* referred to governing authorities that were embryonic by comparison to later 'states' with permanent standing armies. Yet, however under-developed these sixteenth-century regimes were, the oversight of military matters was a central task for them, alongside policing, justice and, of course, tax-raising. The management of firearms cut across these various responsibilities, and guns therefore offer a prism through which we can view aspects of state formation that are often dealt with separately in the literature: the development of the military state with its concern for defence on the one hand, of the disciplining state with its concern for social order on the other, and of the associated 'civilizing process'.[24] I will introduce these briefly in turn.

The idea of the military state emerged from a long-running debate about whether there was a military revolution in Europe.[25] That debate began in the 1950s, when Michael Roberts made the case that the tactical innovations of warfare between 1560 and 1660 led to a growth in the authority of the state, especially in relation to tax-raising. (He referred primarily to complex infantry techniques requiring extensive drill and training.) Subsequent debate focused not so much on whether there had been a military revolution but on its timing, with contributors highlighting the significance of both earlier and later developments.[26] The Italian states were rather marginal to discussions of the military revolution, although Giuseppe Del Torre has made a case that the taxation system introduced by Venice after the War of the League of Cambrai (1508–16) laid the groundwork for that city-state's long-term significance in European politics.[27] In relation to the Italian Wars, further nuance has been added by Idan Sherer's observation that what may not look like a revolution in hindsight can certainly look revolutionary to those involved.[28] In the 1990s, Clifford Rogers attempted to bring together the wider scholarship with the concept of 'punctuated equilibrium', in which the development of the state was characterized by a 'series of intense revolutionary episodes' within a longer evolutionary process.[29] Emphasizing the concept of evolution more strongly yet, some twenty years later Frank Jacob and Gilmar Visoni-Alonzo argued for the abandonment of the military revolution concept altogether. In light of newer scholarship on global history, they particularly criticized the idea that a distinctly western European military revolution had enabled the continent's powers to build their global empires, pointing out that the new methods were not widely used in overseas conquest and that, in any case, military development was well advanced elsewhere in the world.[30]

Even while scholars question the concept of a military revolution, some concepts from this literature remain useful, especially those relating to the two sides of the state: the fiscal state, whose need for a standing army motivates its efforts in tax-raising, and the contractor state, which purchases services. Recent research has built on the concept of fiscal-military/contractor states through a transnational perspective focused on those European cities which from the 1530s onwards became 'fiscal-military hubs'.[31] This work challenges the idea that the state was necessarily the most important organizing factor in military supply. Whether or not one thinks the idea of a military revolution is helpful, the fact that early modern states often contracted out arms purchases is certainly important to understanding developments in the small arms industry.[32] Nor is the idea of a fiscal state irrelevant: effective use of small arms on the battlefield required drilling, which in turn had implications for the organization of military personnel, including the militia that became a vital route for gun proliferation. Training armies in the new techniques had costs that needed to be met through taxation, and guns were more expensive than some (if not all) alternatives.[33] Moreover, there does not need to have been a military revolution for there to have been a 'firearm revolution'. Handguns matter precisely because they escaped the military sphere. Their use expanded over decades and persisted over centuries, as did the debates about how to manage them. Jacob and Visoni-Alonzo are right to observe that 'the development of new military technology does not necessarily have to change society as a whole'.[34] Like print, however, handguns changed human relations. If we can speak of a print revolution in early modern Europe, then we can speak of a firearm revolution too.

This book brings the literature on the military aspects of the state into dialogue with discussions of the role of government in the maintenance of social order. Discipline is, of course, not insignificant in discussions of the fiscal-military state, which emphasize the importance of collective manual drill in disciplining armies.[35] However, it remains rare to see these different state priorities treated in tandem. Yet, if defence was one challenge for the early modern state, addressing crime and disorder was another, and scholars of violence have also taken an interest in the use of firearms. This is a particularly pressing historical question in the Italian context: it is generally agreed that the rate of homicide in sixteenth- and seventeenth-century Italy was relatively high compared to northern Europe.[36] Robert C. Davis's essay on outlaws and their firearms in the late sixteenth-century Papal States addresses some of the attempts to legislate bans on wheel-lock guns in relation to social change and the process of 'refeudalization' in early modern Italy.[37] Moreover, numerous scholars have pointed to the importance of the early modern state in the social disciplining of its people, whether through the development of norms around manners and the rise of civility, through its legal provisions or through surveillance.[38] Recent work, however, has tended to emphasize state-building from below, assessing the role of estates, communities and social conflict in state formation, and the emphasizing the significance of personal agency.[39] Such approaches have tended to highlight the limitations of the state: Joanna Innes suggests that in relation to welfare organization 'early modern states were rarely much more than co-ordinating agents'.[40] This echoes the approach of the latest military histories.

In this book, therefore, I approach the military revolution debate side-on, taking one of its components—the rise of small arms—and assessing its impact on society at large. Beginning with the original, military, function of handguns, I identify the processes through which, despite initial suspicion, they came to be familiar objects in western Europe. The proliferation of guns via militia service made them accessible for use in farming, but also by bandits. Rising crime led to demands to carry guns for self-defence, and although their usefulness for self-defence was questioned, by the time that debate arose they were entrenched as a technology. Their producers had a direct interest in continuing to make them, and the many ancillary industries had an interest in their continued use too. Governments, meanwhile, were concerned to ensure they had adequate capacity to increase production in the all too likely event of war. These intertwined processes, together with a growing cultural acceptance of handguns, together made for a firearm revolution that left western Europe with a lethally armed citizenry and normalized the new technology.

Moreover, thinking about a firearm revolution in this way gives us a holistic perspective on the early modern state. The production of handguns for military use, and the training of shooters, is an important aspect of their history, but it is far from being the only aspect. If, as Robert Barret observed in 1598, 'the wars are much altered since the fierie weapons first came vp', so too was life beyond.[41] As the case of Vincenzo indicates, guns had non-military uses: for livestock control and hunting, and for personal protection. Given the lethality of these weapons, they raised concerns about social order, confirmed by their use in robbery and assassination. They also had a significant role to play in the process of imperial expansion, and the trade in gunpowder weapons was soon a matter of international diplomacy. Finally, and perhaps most importantly, they changed relationships between human beings. As Bruno LaTour observed, 'You are different with a gun in your hand'.[42] With guns, there could be gunmen: the possibility of killing that guns represented made for a revolutionary change in human relations. Indeed, the presence of firearms in society made a civilizing process all the more important. This book brings together these diverse approaches to the history of handguns to produce an integrated picture of their production, use and cultural meanings in sixteenth-century Italy and beyond.

The sources for the history of handguns in sixteenth-century Italy are fragmentary. Western Europe had a notably ambivalent attitude towards firearms, and this results in some significant silences.[43] Still, the archival record is more substantial than it might appear at first sight. In the nineteenth century, several useful compilations of primary sources were produced, which remain important points of reference, but regime change in the large centres of Bologna, Florence and Milan disrupted institutions and record-keeping, and there are few military archives with consistent series through the Italian Wars (that of the Duchy of Ferrara, in Modena, is an important exception; the Siena records, which have been studied by Jacopo Pessina, likewise).[44] The records of Brescia and Venice, which might shed light on handgun production, have numerous lacunae: the earliest sixteenth-century Brescian ducal register covers the years 1528–33, but consistent records begin only in 1546; only a single volume detailing the issue of arms licences in sixteenth-century Bologna, covering four months of 1552, survives.[45] Sometimes, however, the absences have been exaggerated: records do exist, but it is a matter of knowing where to look for them.[46] I was told in one Italian archive that they held no relevant documentation for a study of firearms, only to find guns listed in the records of the ducal armoury. The 'material turn' in history suggested to me—and so it has proven—that the more conventional military and legal sources might well be complemented by evidence from inventories and account books as well as by objects themselves and surviving representations.

Moreover, lacunae in one Italian archive can often be compensated for with documents from elsewhere: this study thus draws on archive material from across north and central Italy, including Rome, Florence, Bologna (second city of the Papal States), Venice and its subject city Brescia, and Modena (where the archive of the Duchy of Ferrara is located); I have further consulted published primary sources from Venice and Brescia as well as those referring to other Italian city-states, including Lucca and Siena. The types of documentation cited include rulers' decrees, account books, inventories, correspondence, judicial records, export licences, arms licences and military surveys. Letters and account books of noble families who acted as *condottieri* (mercenary commanders) and arms brokers show how transactions were conducted in practice, which allows us to establish how firearms supply networks intersected with (or broke from) prior practices of arms supply. Criminal investigations reveal details of firearms offences but also the broader social context of non-elite gun use.

This material is complemented by the study of surviving firearms from numerous museum collections, including those of the Royal Armouries Museum in Leeds, the Museu Militar of Lisbon, the Kunsthistorisches Museum (Vienna), the Metropolitan Museum of New York, the Chicago Art Institute, the Bavarian National Museum, and the surviving armoury of the Este dukes of Ferrara—the only consistently preserved Renaissance armoury—at Konopiště Castle near Prague. I consider visual evidence, including manuscript illuminations, weapon designs, sketches, portraits, tapestries, frescoes, furniture and sculpture, from museums and historic houses including the Villa del Principe in Genoa and Palazzo Vecchio in Florence, the Museo del Prado, National Gallery of Ireland and Tate Britain. In a number of cases, the focus on firearms allows me to add nuance to existing interpretations of these artworks, and to identify some key innovations, notably in relation to the likely earliest sculptural representation of European guns. I also assess the absence of firearms from representations of places where we know they were present, asking why, for example, did Titian paint the Emperor Charles V with a gun only for Van Dyck to leave the weapon out when he produced his own version of the portrait? Taken together, this rich range of sources illustrates the production, distribution and maintenance of firearms, modes of use, the routes by which guns proliferated, the mechanisms of regulation that states attempted to impose, the variety of loopholes and flaws in the regulatory system, and, of course, wider cultural perceptions that shaped social attitudes towards the gun.

The book is divided into three parts, moving from a deep local case-study of arms production and contracting to wider issues of gun ownership and gun control in Northern Italy, and finally to Europe and its empires. Part I addresses handgun proliferation in military contexts, assessing how guns came to be adopted, how they were produced and distributed, and the social implications of these processes. Chapter 1 outlines the development of portable firearms and their changing use in warfare, situating the Italian Wars (1494–1559) in international perspective and setting the scene for the discussions that follow. Chapter 2 focuses on the world of arms manufacturers. It investigates the processes by which an arms industry developed and gained political leverage in northern Italy, focusing on Brescia, a subject city of Venice, and the small valley town of Gardone Val Trompia in its hinterland, which in the sixteenth century became a major centre for arms manufacture and remains so to this day. Chapter 3 then sets the process of military contracting in wider social context through a case-study of a Brescian patrician, Giovanni Battista Porcellaga, who acted as broker for arms purchases, and through consideration of wider distribution networks. As a major theatre of European conflict in this period, Italy offers a rich case-study in the social dynamics of gun production and use, and this microhistorical study helps us understand precisely how firearms were assimilated into society. Moreover, while these chapters have a local focus, this is a global story too: guns from Gardone Val Trompia were found on the wreck of the *Mary Rose*, which sank off southern England in 1545, while Cesare de' Fedrici, a man from the arms-producing valleys, travelled to what is now Myanmar in 1569 and reported that there he had seen arquebusiers drilling daily.[47]

Part II turns to investigate firearms in the civilian context and attempts to regulate them across north and central Italy. Chapter 4 addresses court culture and luxury guns, exploring the changing types of weapons owned over time, and examining the imagery that decorated guns, which ranged from the religious to the erotic. It demonstrates how—through the integration of guns into existing cultures of gift-giving, and through the use of familiar designs and materials—firearms were assimilated into the cosmopolitan world of the court, with technologies exchanged across Europe. Chapter 5 turns to wider civic gun cultures, exploring how the development of militia helped familiarize the men of the north and central Italian states with guns and drawing on qualitative material from legal and criminal trial records to analyse firearm use by people of different rank and gender. Chapter 6 then assesses the attempts at gun control enacted on the Italian peninsula through the course of the sixteenth century, investigating how the many limits and exemptions to the legislation were justified.

Part III of the book looks beyond the Italian peninsula, towards the final thirty years of the sixteenth century and into the seventeenth. Chapter 7 explores the use of firearms in a series of attempted and successful assassinations between the 1530s and 1580s, ranging from northern Italy to the Netherlands via England and France, including the well-known assassinations of the earl of Moray, regent of Scotland in 1570, and William the Silent in 1584. The case-studies shed light on cultural attitudes, technological capabilities, and the continuing debate about firearm proliferation in the last quarter of the sixteenth century. Chapter 8 turns to consider representations of firearms in European visual culture, from the early Amsterdam militia portraits to the Catholic iconography of the Battle of Lepanto, to the French Wars of Religion and Dutch Revolt. Both chapters draw contrasts and comparisons between the Italian material and northern Europe. Chapter 9 then turns to the rising European empires and to new representations of the gun in imperial and colonial contexts. Taking as its starting point the collection of *Navigations and Voyages* published by Giovanni Battista Ramusio in Venice from 1550, and continuing through a study of the visual culture of firearms, this chapter investigates the association of guns with increasingly racialized ideas of European technological superiority, setting the scene for the coming centuries in which the arms trade became fundamental to the functioning of empire. While the Italian states themselves were rarely directly involved in these new colonial projects, numerous individual Italians and private firms engaged in global trade via the growing Spanish and Portuguese Empires. Venice was a key centre in the provision of information about these early encounters, while Genoa acted as a financial hub for Spain. Here, as with the Italian Wars, the Italian case-study is one that illuminates societies well beyond the peninsula in a pivotal period for global history, the legacies of which remain with us today.

PART I

The Military Context

CHAPTER ONE

The Italian Wars

THE MOST ICONIC battle of the Italian Wars (1494–1559) is that of Pavia, which took place on 24 February 1525. A commemorative tapestry sequence produced for the Spanish victors provides some of the most detailed visual evidence for the use of guns in early modern European conflict (Figure 1.1). Pavia, some twenty miles south of Milan, had been under siege by the French since the previous autumn, when the Spanish army had been forced to pull back. They had left a garrison of around six thousand men inside, along with many residents. Outside the walls, the French waited them out. Reinforcements arrived late in the day, bearing the firearms that by now characterized the Spanish army; they not only defeated the French but captured their king, Francis I. Not particularly notable in terms of tactics, on a cultural level Pavia was highly significant. While on the one hand the capture was deeply humiliating for Francis, in some accounts of the battle the French emphasized their king's chivalry in contrast to a victory achieved only with dishonourable firearms.[1] Blaise de Monluc, a French commander who as a young man was captured at Pavia, later wished that 'this unhappy weapon had never been devised', complaining that the men who used it were 'often cowards and shirkers, who would never dare look in the eye those who from afar they topple with their wretched bullets'.[2]

On the other hand, firearms are quite apparent in visual depictions of the event. Rupert Heller's painting in Sweden's Nationalmuseum shows a group of arquebusiers in the foreground.[3] They stand out in white overshirts, a device used to ensure they could

FIGURE 1.1 Bernard van Orley, *The Battle of Pavia: Advance of the Imperial Army*, c. 1525–31. Tapestry: wool, silk, silver and gold. Capodimonte Museum, IGMN 144483. © Museo e Real Bosco di Capodimonte, Naples / Bridgeman Images.

identify one another in low light. The commemorative tapestry sequence, produced to a design by Bernard van Orley and now at the Museo di Capodimonte in Naples, features substantial numbers of arquebusiers, including in the depiction of the French defeat two in the foreground, one of them reloading (Figure 1.2). One of these soldiers has a small cross, presumably on a chain around his neck, hanging over his shoulder, an indication that guns were by no means perceived as incompatible with Christian virtue. Behind them, as a line of shooters fires, smoke begins to fill the battlefield. Those with firearms in this image are the infantry: none of the commanders referred to by name on the tapestries is carrying a gun, a reflection perhaps of continuing ambivalence towards the technology, perhaps simply of the fact that designs suitable for use on horseback were at the time of the victory still rare. On the tomb monument of Francis I, the defeated party at Pavia (located in the

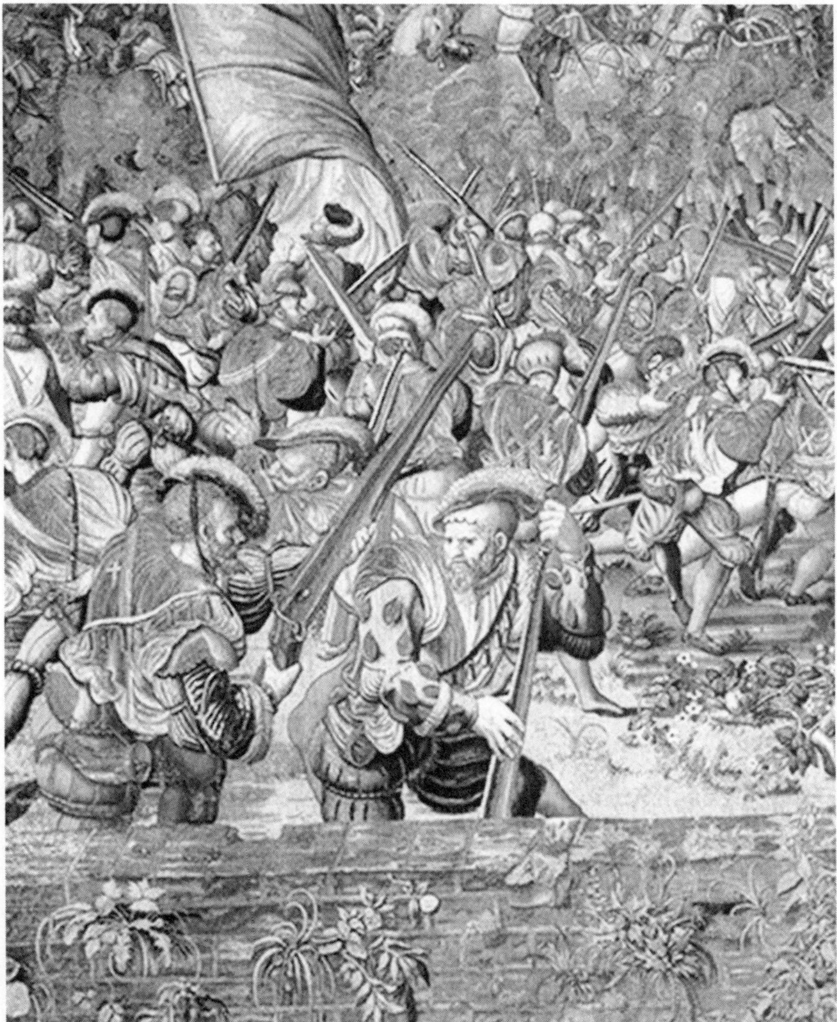

FIGURE 1.2 Bernard van Orley, *The Battle of Pavia: Defeat of the French Cavalry* (detail), c. 1525–31. Tapestry: wool, silk, silver and gold. Capodimonte Museum, IGMN 144484. Museo e Real Bosco di Capodimonte, Naples. Photo: public domain via Alamy.

Basilica of Saint Denis), arquebusiers are less prominent: one features in the depiction of the 1515 Battle of Marignano, but cavalry, apparently without firearms, dominate the relief showing France's 1544 victory at the Battle of Ceresole.[4] Yet, by the time of Francis's death in 1547 they were increasingly important on the battlefield.

The following year, Titian painted an equestrian portrait of Francis's great rival Charles V, Holy Roman Emperor. This is, in all likelihood, the first individual European portrait to show its subject with a gun.

This chapter documents the growing use of firearms during the Italian Wars (1494–1559), setting the scene for more focused discussions of the arms industry and arms brokers in the subsequent chapters of this part, and for the assessment of guns' impact beyond the battlefield in part II. It begins with an overview of gunpowder technology, outlining its development from its origins in twelfth-century China to the European battles of the early sixteenth century in which handguns became decisive. It sets out the political context for the Italian Wars, pausing to assess key moments in the conflict (including the 1512 Sack of Brescia and the 1527 Sack of Rome) that illuminate practices of warfare; it further introduces some of the contemporary critiques, including those of Leonardo da Vinci, whose notebooks include numerous designs for weapons. Of particular interest are the shifting experience of the battlefield in terms of sight and sound; the changes to medical practice; and the need for military discipline and drill that the new technology necessitated.

The development of the handgun

Gunpowder technology originated in China: sculptural evidence in Sichuan suggests guns may have been in use in the first half of the twelfth century and a surviving Chinese gun can be dated no later than 1288.[5] Debate continues on why China subsequently lost the initiative in its development to Europe: Kenneth Chase noted that after 1368 China's wars were primarily conducted on horseback, on the steppes, an environment in which the available gunpowder technology was impractical, while Tonio Andrade argued that this could only be one part of the story, pointing to continuing infantry warfare to China's south and the lull from the mid-fifteenth century.[6] (It was well into the sixteenth century before firearms suitable for cavalry use were developed.) The Mongols had some types of gunpowder device as early as the 1230s, and are the most likely source of the transmission of the technology to Europe.[7]

Firearms may have been used at the Siege of Huesca (north-eastern Spain) in 1324, although it is not certain; they were recorded in both England and Florence in 1326 and were known in Russia by the 1380s.[8] The Ottoman Empire had gunpowder weapons in the same period; the Mamluk Empire may well have got there earlier, deploying cannon from the 1360s.[9] Although there is a possibility that firearms technology reached the Ottoman Empire from Central Asia, the key transmission route in the fourteenth century was via the Balkans, in the context of conflict with Venice; imported Genoese guns were also documented in Constantinople before the end of the century.[10] The evidence of the 1453 conquest of that city is that the Ottomans were able to deploy the new technology to considerable effect.[11]

Gunpowder technology thus established, something of an arms race ensued. Cannon were initially used principally for siege warfare and work took place both to improve their power and their portability, so that they might be used more effectively on the battlefield. In Christian Europe, techniques from church bell manufacture were adapted to cast muzzle-loading bombards in bronze.[12] In the middle of the fifteenth century, French gunners switched from using stone balls to cast-iron shot: this 'allowed the calibre and consequently the size, weight, and cost of guns to be reduced, while at the same time achieving considerable improvement in shot penetration'.[13] Florentine gunmakers were also experimenting with the use of bronze.[14]

The fifteenth century saw the first substantial proliferation of something approaching the modern handgun through Europe. Small portable guns were known in Italy by the second half of the fourteenth century,[15] but recent research has credited the first arrival of long wood-stocked handguns in Italy to Sigismund of Luxembourg, who travelled in 1431 to Milan for his coronation as king of Italy, and then in 1433 to Rome, where he was crowned Holy Roman Emperor. Sigismund brought with him an entourage including six hundred 'iscoppettieri colle bonbardette' (handgunners with little bombards): these weapons were produced in Nuremberg and were the first of their kind to be seen in Italy.[16] Civic defence ordinances for the German lands were already referring to firearms in the late fourteenth century, and Pope Pius II (r. 1458–64) wrote in his *Commentaries* that the 'scoppetum' was 'a weapon invented in Germany in our time'.[17] Over time the more basic guns, which did not have a trigger, were refined by the addition of a matchlock mechanism, which enabled the shooter to hold the gun with both hands while firing rather than having to lift one hand to light the powder.[18] Relatively few examples of such plain guns survive: most such weapons have rusted away, with only prestige examples being collected. One dating from around 1500 may be found in the collections of the Kunsthistorisches Museum, Vienna.[19] A somewhat later, but similarly plain, weapon, manufactured around 1540 in Brescia, is in the collection of the Royal Armouries in Leeds. This gun is 130 centimetres long; similar weapons were found on the *Mary Rose*, an English warship that sank in 1545.[20]

Small arms were quickly adopted in the decades following Sigismund's journey to Italy: a Florentine *cassone* (storage chest) panel from the 1460s, now in the National Gallery of Ireland, Dublin (Figure 1.3), shows handguns with wooden stocks in use at the Battle of Anghiari, which took place in 1440; these were soon in use outside military contexts and a handheld firearm (described as *una spingarda de ferro*) was used in a homicide in Bologna that same decade.[21] The Dublin panel is notable not only as one of the earliest visual depictions of Italian wartime firearm use, but also because *cassoni* were widely used in the Italian domestic interior and such a portrayal in the home environment would serve to normalize the technology.[22] Guns, meanwhile, were increasingly deployed in military contexts. Troops were armed with handguns at the 1448 Battle of Caravaggio between Venice and Milan; no fewer than fourteen thousand were among the ordnance offered in 1463 to Pope Pius II by the Venetian authorities, to be used in his planned crusade.[23] A Niccolò d'Este (probably the claimant to the marquisate of Ferrara) bought fifty *schioppetti* in 1469, and in 1483 the castellan of Casale sul Po described the fortress as 'well-supplied with bombards and arquebuses'.[24] A 1490s inventory of the fortress at Valiano (near Montepulciano in Tuscany) included a variety of firearms, among them fourteen *schoppietti* 'between good and worn-out' (*fra buoni e tristi*).[25]

This growth in the use of small arms was matched elsewhere in Europe and around the Mediterranean. The Mamluk Empire likewise had handguns in the fifteenth century; the Ottomans

FIGURE 1.3 Unknown artist, *The Battle of Anghiari* (detail), late 1460s. Tempera and gold leaf on poplar panel. 62.4 × 207.5 cm. Bequeathed, Sir Hugh Lane, 1918. National Gallery of Ireland, NGI.778. Public domain.

had them from at least 1465, although a source from 1510 suggests that at that point the elite troops known as Janissaries had only recently learned how to use them.[26] This seems unlikely, however, because already in the 1470s the Venetians were actively supplying their Safavid allies with handguns to assist them in opening up an eastern front against the Ottomans, whom the Venetians were fighting from the west.[27] The Pastrana Tapestries (commissioned by the Portuguese king Alfonso V to celebrate his 1471 conquest of Asilah and Tangiers in Morocco and produced by Flemish weavers in Tournai) feature guns in use on both sides of the conflict. The majority of the firearms shown are rudimentary handguns without a trigger mechanism (Figure 1.4), but more advanced matchlocks are also shown alongside crossbows and artillery.[28] Guns had become decisive in battle by the early sixteenth century, and were important in both the Italian Wars and the Ottoman-Mamluk conflict that culminated in Ottoman victory at the Battle of Marj Dābiq in 1516. Under Ivan the Terrible, the Russians had musketeers by the middle of the sixteenth century,[29] but a Genoese visitor had observed handguns in use in Circassia, on the north-eastern shores of the Black Sea, by 1502.[30]

FIGURE 1.4 Nuno Gonçalves (design) and attr. Pasquier Grenier (manufacture), *The Siege of Asilah* (detail), Tournai, 1471. Wool and silk, 11×4 m approx. Colegiada de Pastrana, Guadalajara. Photo: public domain via Alamy.

Terminology and capability

The terminology of small arms was not precise in this period, but in the sixteenth-century context, 'arquebus' typically referred to a gun that could be fired from the shoulder without a support to steady it.[31] An *archibusetto* was a smaller version. The *archibusone* (literally, 'large arquebus') required a stand and is generally listed alongside heavier artillery in the records. Handguns were also known as *schioppi* (also spelled *stioppi*) or in smaller sizes *schioppetti*; this weapon is sometimes (but not necessarily) distinguished from the arquebus by the fact that an arquebus has a trigger while a *schioppetto* requires both hands to fire.[32] In practice, however, usage was not consistent and may have been affected by considerations of literary style as well as regional preference. I have therefore chosen to translate *archibugio*/*archibuso* to 'arquebus' but to leave *schioppo* (plural *schioppi*) in the Italian, standardized to that spelling for ease of searching, so that those readers with an interest in the detail can see the linguistic nuance.

Guns came in three sections, which were produced, and could be bought and sold, separately: a lock (the firing mechanism), a stock (the wooden case), and a barrel. The first and last of these required specialist metalworking skills, but wooden stocks were straightforward to produce. Military guns were most often fired with what was called a matchlock, which brought a long, slow-burning cord—the match—into contact with the gunpowder (Figure 1.5). An alternative mechanism, the wheellock, was developed late in the fifteenth century and became a source of significant social anxiety because it could be concealed beneath clothing, a question to which I return in part II of this book. Most surviving wheellock pistols are highly decorated examples, but relatively plain weapons of this type can be found in several European collections (Figure 1.6).[33] (Morin suggests that one example in Madrid, which bears the mark of Peter Peck, may in fact have been used by Charles V for military purposes, his more elaborately decorated version reserved for the parade ground.)[34] Most soldiers, however, continued to use matchlocks, with the exception of elite cavalry for whom the advantage of firing one-handed outweighed the cost and reliability problems associated with the wheellock.

A number of accessories were also required to use a gun: a large powder-flask in which to store gunpowder and a small priming-flask with which to measure the specific dose required for the weapon. Powder, shot and match-cord were needed too. A ramrod was used to push the shot down the barrel; this was often stored parallel to the barrel in a hollow stock. The arquebusiers in the Van Orley tapestry are shown without armour, wearing the flamboyant slashed doublet and hose typical of Landsknechts (freelance soldiers from the German lands). The orders and export licences for firearms discussed in chapters 2 and 3, however, show that when guns were purchased for military use, orders frequently included a morion (helmet typically worn by an arquebusier) and corselet (breastplate). These were less protective than full armour but allowed for greater manoeuvrability.

Tests on the accuracy of early modern firearms undertaken in the 1990s with examples from the historic armoury at Graz provide important evidence for their efficacy. Most guns in this period were not rifled, which significantly limited their accuracy. Rifling—that is, the incorporation of a spiral design into the inside of the barrel, thereby causing the shot to spin and in turn improving

FIGURE 1.5 Attrib. Jacques de Gheyn II after Hendrick Goltzius, *Soldier with Arquebus*, 1587. Engraving on laid paper, 35.4×25.2 cm. National Gallery of Art, New Century Fund, 2004.8.7. Public domain.

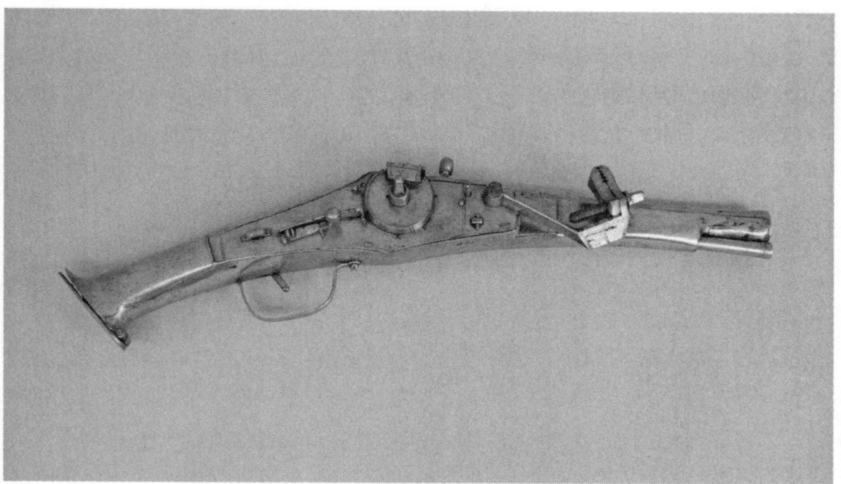

FIGURE 1.6 Unknown maker, wheellock pistol, Nuremberg, 1579. Iron, chiselled and etched. 460 × 80 mm. Victoria and Albert Museum, M.628-1927. © Victoria and Albert Museum.

accuracy—was probably developed around the turn of the fifteenth to the sixteenth century.[35] A rifled matchlock from 1560 is in the Fitzwilliam Museum, and another from 1581 in the Metropolitan Museum of Art.[36] However, the technique was not widely adopted until the seventeenth century, and routine military use came even later.[37] In test conditions (in which the guns were sighted and placed on a rest, eliminating possibilities for human error), a torso-size target at a distance of 100 metres was hit around 50 per cent of the time with more conventional sixteenth-century firearms.[38] Furthermore, the velocity of the shot declined significantly at range, meaning that the potential damage done by these weapons was reduced—from a 369-cubic-centimetre cavity at 8.5 metres to a 155-cubic-centimetre cavity at 100 metres. In some cases, plate armour proved entirely protective. On the other hand, the tests confirmed that at the closer ranges the physical damage from a shot was considerable. It was not implausible for the brother of Benvenuto Cellini, author of an important sixteenth-century memoir (to which we will return) to believe on being shot in the leg that he had only 'a few hours left'.[39] The theorist of war Raimond de Beccarie de Pavie, baron de Fourquevaux, observed in his 1548 *Instructions for the Warres* that 'any soldier of mean judgement' would regard it as ill-advised not to wear armour given that it afforded at least some protection from other types of weapon and from guns that were poorly charged, overheated or shot from a distance

> for although that Harnes [armour] be too weake to resist ordnance or Harquebushes; notwithstanding, it dooth defend a man from the stroke, of Pike, Halbard, and Sword, Crosse-bowe, Long-bowe, and from Stones, and from all other hurt, that may proceede from the enemies hande, and sometimes a Harquebuze may bee so ill charged, or so hotte, or may bee shotte so farre of, that a Harnes if it be good, may saue a mans life.[40]

In the battlefield environment, the limited accuracy of weapons was compensated for by massed ranks of shooters, who acted as a sort of human machine gun in the 'pike-and-shot' formations that became dominant in this period. Even off the battlefield, a 50 per cent chance of receiving such a lethal wound (far larger than could be made by a dagger or arrow), undoubtedly had an intimidatory effect. It is worth remembering that in a successful hold-up, nobody gets shot: the precise accuracy of the gun matters less than the calculation of whether its holder is prepared to shoot. Cellini described his own successful use of this technique, in which he used a loaded hunting gun to defend himself against some Spaniards who were threatening to beat him up after he had fallen out with one of their countrymen over an art commission.[41] In the battlefield context, the devastation that could be wreaked by mass shooting led many commanders to avoid pitched battles.[42]

The lack of accuracy can also be overstated. Certainly, by the 1530s guns were in use for hunting. Cellini used one for fowling, which requires sufficient precision to hit a moving target; a 1544 gun belonging to Duke Cosimo de' Medici of Florence (discussed in chapter 4) incorporates a design of hunters using a handgun for this purpose.[43] These were elite users with access to good quality weapons, but by mid-century farmers too were using guns, which implies that even beyond the high-end market other advantages of the weapon were perceived to outweigh its restricted accuracy.

From artillery to handguns

Against the backdrop of a growing, if often ambivalent, interest in small arms technology, in 1494 a dispute between two members of the ruling house of Milan triggered a conflict on the Italian peninsula that would last for more than six decades.[44] Italy at the time was divided into five large states (Milan, Florence, Venice, the Papal States, and Naples), and many smaller ones, including Ferrara, Mantua and Urbino, whose rulers were frequently contracted to lead mercenary companies on behalf of the larger polities. Ludovico Sforza, who had been governing Milan during the minority of his nephew Gian Galeazzo, refused to hand over power and solicited the support of Charles VIII, king of France, against the nephew, who in turn was backed by his father-in-law, Alfonso II, king of Naples. Charles VIII saw this as an opportunity to assert the old Angevin claim to rule in southern Italy, and marched troops not just to Milan but down the peninsula. The Spanish monarchs soon entered the fray, and there followed a series of wars that drew in multiple states in a variety of alliances, and which became a testing-ground for new military techniques.

The initial innovation was not so much the handgun but the heavier artillery. By the time of the French descent into Italy and the start of the Italian Wars in 1494, their cannon, made of bronze rather than iron, were, as historian Francesco Guicciardini recorded some decades later,

> smaller without comparison than those of stone made use of heretofore, and [they] drove them on carriages with horses, not with oxen, as was the custom in Italy; and they were attended by such clever men, and on such instruments appointed for the purpose that they almost ever kept pace with the army.[45]

The weapons were now just light enough that teams of Swiss mountaineers could drag them over the Alps.[46] They became increasingly viable for use on the battlefield, notably at the Battles of Ravenna (1512) and Marignano (1515).[47] While technology was by no means the only decisive factor in any given conflict—holding out when besieged depended on supplies and, eventually, relief—it was undoubtedly significant.[48] The challenge for those facing the improved cannon was to develop defences that could protect their cities from the new threat. Early experiments with such defences were often relatively simple, involving new earthworks or an increased use of artillery from inside besieged towns. The most significant development in defensive technology, however, was what became known as the *trace italienne*. Relying on lower, thicker walls with ditches either side, rather than the high fragile walls vulnerable to cannon, the star-shaped forts characteristic of the *trace italienne* enabled fire to be directed against anyone trying to scale their walls.[49] This shifted the balance of power in any given conflict in favour of cities trying to protect themselves from attack.[50]

Early interest in these types of firearms is evident in the notebooks of Leonardo da Vinci (1452–1519). Although Leonardo is now predominantly known for his art, he was also a sought-after military engineer. In the 1480s, pitching his services to the duke of Milan, he wrote: 'I have also plans of mortars most convenient and easy to carry with which to hurl small stones in the manner almost of a storm; and with the smoke of this cause great terror to the enemy and great loss and confusion.'[51] An image of such weapons appears in his Atlantic Codex.[52] Here is evidence not only that the potential to evoke terror was a key element of what a designer of armaments could offer, but that this featured in the promotion of services. While artillery dominates in his notebooks, as would be expected given their dating, around 1485–86 Leonardo drew a number of smaller firearms, including examples of arquebuses and *schioppetti*, and devices for firing a number of *schioppetti* at once (Figure 1.7).[53] There is no direct evidence that his more adventurous multi-barrelled gun designs were produced in his lifetime, but there are some examples from later in the century. In 1560, one man was exiled from Venetian territory after producing a four-barrelled gun for the purposes of assassinating a local nobleman and his brothers: according to the letter reporting his case (which also noted various prior crimes) this was a 'diabolical instrument'.[54] A fourteen-barrelled example survives in the collections of the Museu Militar in Lisbon, while a 1578 inventory of goods belonging to Paolo Giordano Orsini, duke of Bracciano (in the countryside north of Rome), refers to part of a twelve-barrel arquebus.[55] Paolo Giordano served the king of Spain as a general, so an

FIGURE 1.7 Leonardo da Vinci, Codex Atlanticus, c. 1485, fol. 977r. Veneranda Biblioteca e Pinacoteca Ambrosiana, Milan. © Veneranda Biblioteca Ambrosiana / Mondadori Portfolio.

interest in military technology might be expected. Examples of luxury double- and triple-barrelled guns from the sixteenth century do survive in larger numbers, as we will see in chapter 4.

Yet Leonardo's ambivalence about gunpowder weapons is also clear. In multiple prophecies and observations (drawing on understandings of iron and mining that dated back to the ancient world), he wrote of the destruction guns might wreak, describing the iron from which they were produced as a 'monster': 'How much better were it for men that thou shouldst return to your cave!'[56] Leonardo gave an even more emotionally evocative account in his prophecy 'Of Metals':

> These shall come forth out of dark and gloomy caves; that which will put the whole human race in great anxiety, peril, and death. To many that follow it, after many sorrows it will give delight, but whosoever does not side with it will die in want and misfortune.[57]

Whether Leonardo is speaking of money or weapons is ambiguous, perhaps deliberately so. They are linked in his prophecies by their common emergence from below ground. In a prophecy on money and gold, Leonardo described how 'out of cavernous pits a thing shall come forth which will make all the nations of the world toil and sweat with the greatest torments, anxiety and labor, that they may gain its aid'.[58] Discussing 'great guns, which come out of the pit and mold', he described 'that which with its terrific noise will stun all who are near and with its breath will kill men and destroy cities and castles'.[59] It is clear that in relation to metal, money and weapons, the associated emotions for Leonardo were anxiety, sorrow and terror. Their underground origins also associate them more subliminally with the classical Underworld or Christian Inferno, and thus with another contemporary idea about guns: that they were, to quote Erasmus of Rotterdam (a prominent Catholic reform thinker) on cannon, 'weapons invented in hell'.[60]

Indeed, the Book of Hours of Charles d'Angoulême (father of Marguerite of Navarre and Francis I of France), illuminated by Robinet Testard sometime between 1480 and 1496, provides a visual representation of this idea. A miniature of the Resurrection shows, in one corner, a pair of demons, one armed with a handgun, shooting at the risen Christ from a balcony (Figure 1.8). A frontispiece

FIGURE 1.8 Robinet Testard, *Hours of Charles of Angoulême*, 1480–1496, fol. 110v (detail). Bibliothèque Nationale de France / Gallica. Public domain.

for Augustine's *The City of God*, produced in Basel and redrawn in Venice in 1489–90, likewise shows a demon, this time aiming at the godly city.[61] This echoed earlier imagery associating crossbows with demons.[62] Indeed, attempting to explain the outbreak of the Great Pox during the 1490s, Coradino Gilino, court physician at Ferrara, looked to godly wrath at 'impious deeds', and went on to describe a 'most terrible distemper that is raging not only in Italy but throughout the whole of Christendom':

> Everywhere the blare of trumpets is sounding, everywhere is heard the clash of arms, everywhere are being constructed military weapons, bombards, instruments and a great many engines of war, moreover

> instead of the spherical stones, which have been in use up to the present time they are now making iron balls, a hitherto unheard-of thing.[63]

While in general Christian attitudes towards firearms were ambivalent, this had also been true of attitudes towards earlier missile weapons. These had been banned outright in 1139, but the anathema was 'almost universally ignored' and cited primarily when diplomatically convenient. Papal armies certainly included crossbowmen.[64] In practice, questions of any given weapon's acceptability were subsumed into wider discussions about whether the war in question was just.

The rise of the arquebusiers

In parallel to developments in artillery, handguns were increasingly adopted for military use. The militia in Friuli, where Venice bordered Ottoman territory, were equipped with handguns from the 1490s, and made important if irregular contributions to Venetian defence.[65] It was, however, Spanish units that made the running in the use of firearms in the major battles of the Italian Wars.[66] The first battle in which effective use of guns was decisive was probably that of Cerignola (in the Realm of Naples) in 1503, where a combination of earthworks and arquebus fire enabled the Spanish infantry to rout the French.[67] Spanish and German instructors were hired to train the Florentine militia, who were required to drill with firearms from 1508.[68] (The militia were an important route for transmission of shooting skills, a question I discuss further in chapter 5.) The German mercenaries known as Landsknechts were also prominent among gun users in the early stages of the Italian Wars, and in the visual culture of the German lands, arquebusiers are portrayed in greater detail and at an earlier date than in Italy.[69] A late fifteenth-century hand-coloured woodcut, now in a private collection, for example, shows soldiers firing matchlock arquebuses; an early sixteenth-century series of engravings titled *Triumph of the Emperor Maximilian I* features arquebusiers in the first of three groups of 'valorous soldiers' (the others have large swords and lances) (Figure 1.9).[70] The *Book of Armaments*, an illustrated manuscript that describes the contents

FIGURE 1.9 Arquebusiers, from Hans Burgkmair, *Triumph of the Emperor Maximilian I*, Augsburg, 1512–19. Woodcut on paper, approx. 38 × 38 cm. Victoria and Albert Museum, 13079.119. © Victoria and Albert Museum.

of the Emperor's armouries, includes a range of handguns.[71] (This was compiled after 1515 but initial work was done more than a decade earlier.)

By the turn of the fifteenth to the sixteenth century, however, handgun skills were well established in at least parts of Italy. The *schioppettieri* in the duke of Urbino's service in 1505 were predominantly from the surrounding region.[72] In 1507, the Venetian officials in Brescia reported that there were three hundred men well equipped with arquebuses and *schioppetti* in the comune of Bagolino in Valsabbia (near Lago d'Idro in the Italian Lakes area).[73]

Firearms are also listed in substantial numbers in the fortresses of the duke of Ferrara in 1509,[74] although that same year Pope Julius II (1443–1513) hired three hundred Swiss *schioppettieri* for his army,[75] suggesting that a case could still be made for importing expertise.

Guns had significant advantages. They may well have been cheaper than the crossbow and certainly required less physical strength to shoot than the longbow.[76] The main issue to be addressed was the weight of the weapon. (Here there is a trade-off between portability and recoil, in that a heavier gun absorbs more recoil than a lighter one.) A 1540s example from Gardone Val Trompia, now in the Royal Armouries,[77] weighs three kilograms, but a 1537 list of iron in the Orsini archives which (unusually) lists the weights of guns, values given included totals of 315 *libbre* for fourteen arquebuses and 259 *libbre* for nine. Assuming a conversion of 303 grams to the *libbra*, this gives an average weight across the twenty-three weapons of just under 25 *libbre*, or about 7.5 kilograms.[78] Although these are listed as *archibusi* and not *archibusoni*, it is possible that they were in fact the larger weapon and would have been operated on a stand. Either way, in a society where the majority of people were much more accustomed to physical work than in present-day Europe, it is fair to assume that most able-bodied adults would have been able to use a gun without too much difficulty.

The principal military use of portable firearms was among the infantry, where they were deployed in combination with pikes in the pike-and-shot formations. Effective use of firearms required rigorous discipline among shooters to enable reloading (not a fast process) during battle. If anywhere on the Italian peninsula ought to have been at the vanguard of firearm use, it was Brescia, close to the major production centre of Valtrompia, which I discuss in chapter 2. Brescia had been part of the Venetian Empire since 1426, but during a wider war in 1512 it was first conquered and then—following a failed rebellion—sacked by French troops. The many descriptions of the sack provide important evidence for the extent of gun use and the military discipline required for this to be effective. Cesare d'Anselmi, a Bolognese observer, explained how the French commander Gaston de Foix marched 'a great company of arquebusiers' down the hill from the city's fortress, firing 'with excellent order' at the citizens:

> Because every time they wanted to fire, at the signal of a voice placed among them, the men on the flank would all lie down on the ground and, the arquebuses being discharged, would immediately get up again. And in this way, bit by bit they penetrated to the foot of the mountain, not however without damage to themselves from the artillery.[79]

In situations where multiple lines of shooters were available, this manoeuvre could be extended such that the front line of the formation would shoot then kneel down to reload, while the next line shot over their heads; the process was repeated until the back line had fired, by which time those at the front would be prepared to start again. As we will see, guns were sometimes objected to on the grounds that they allowed even untrained fighters to kill. Yet while in one sense this was true, the victories most easily attributable to small arms were won by highly trained units like these.

Corroboration for the textual account is to be found in the Van Orley tapestries depicting the Battle of Pavia which, in the absence of surviving weapons, provide some of the best visual evidence for handguns in routine military use in the 1520s. The level of detail included—match-cord, serpentines and triggers are all shown in close-up—along with the variation between weapons suggests a high degree of familiarity with battlefield practice. Produced by 1531 as a gift for Charles V, most likely from the States General (ruling assembly of the Low Countries), the seven pieces show in sequence the events leading up to the Spanish victory at the 1525 Battle of Pavia, in which Francis I, king of France, was captured.[80] The visible firearms are all matchlocks; in several cases match-cord can be seen wound around the stock of the gun or around the arm of its user. The arquebusiers wear powder horns at the hip; some also carry a small priming-flask; some have prepared portions of powder slung on a belt over one shoulder. The stocks of their guns are square and boxy (the Italian word for stock, *cassa*, can also be translated as 'box'); they incorporate a slot alongside the barrel to hold a ramrod; one can be seen in use in the fourth tapestry (the defeat of the French cavalry), where in the foreground a soldier is reloading. The third tapestry in the sequence, illustrating the advance of the imperial army, shows a substantial group of arquebusiers making their way through a forest; the dynamic visual narrative of Van Orley's composition shows guns slowly being lowered as the viewer reads the image from left to right; the final weapons are held at the shoulder in firing position (Figure 1.1). The men are accompanied by a drummer, whose beat along with trumpet signals would have ensured coordinated firing. The tapestries also illustrate the quantity of smoke generated by guns on the battlefield, as well as some very large sparks from the firing pans, which hint at the risks of misfiring, while an example of misfiring is shown in the fourth tapestry. The dangers of gunpowder warfare are similarly apparent in the exploding powder barrel in the fifth tapestry of the sequence. While later in the century we will encounter lone gunmen in growing numbers, in the 1520s, the gun remained predominantly a collective weapon, not so much the province of individuals but requiring a disciplined body of troops.

The Sack of Brescia, 1512

Returning to the previous decade, by 1512 it was evident that at least around the Brescian area firearms were in wide use. It is notable, however, that this detail is more apparent in the local sources (and the more popular verse) than in the literary sources and the histories produced elsewhere, which may reflect greater local familiarity with weapons, elite ambivalence, or both.[81] Pandolfo Nassino, a Brescian nobleman loyal to Venice,[82] accompanied his friend Valerio Paitone, a minor lord in the Valle del Garza slightly north of Brescia,[83] during the events of the sack and wrote up his observations. His description of Paitone underlines the quantity of firearms that might be present at even a small fortress in this area: 'He keeps a most beautiful court at Navi, for the majority of the time, in a place called Monticolo; there he keeps fourteen falconets and spingards and ninety-six arquebuses and a lot of munitions.' Munitions here should be understood to mean supplies generally, not only ammunition. Yet while Paitone might have been amply supplied, this was not universally the case. Nassino noted that things might have gone better in Brescia: as things stood there was only enough powder for each *schioppettiere* to fire once or twice.[84]

Among the verse treatments of the conflict is that of Niccolò degli Agostini, *The Events of the Italian Wars*, published in 1521 and covering events between 1509 and that year.[85] Niccolò was a Venetian subject and his account, written in straightforward and sometimes stereotypical language, includes a number of incidents not included in other treatments of the sack.[86] Among them are multiple references to firearms, including in the hands of the French. 'Bold and ready for the task,' he wrote, 'with arquebuses, darts and arrows, they [the French] caused those in the camp a thousand harms.' Emerging from the city fortress, 'some shot crossbows, some *schioppetti*, some sulphur and fire with miraculous care, some unleashed spingards and cannon, throwing many dead to the ground'.[87] While this account does not offer the detail of tactics provided by Anselmi, it does illustrate the wider impact of missile weapons, and certainly conveys the violence of events. A second verse account, the anonymous *Book or True Chronicle of All the Wars of Italy*, published in 1522, also notes the presence of firearms, but with a somewhat different tone. In this case, the guns cannot be fired due to rain (not entirely surprising, given it was February), and in the narrative this becomes a key factor in the Brescian defeat: 'And a cruel brawl began between them / Some with *schioppetti* and some short pikes . . . / And for the rain they could not fire their guns / The French began a cruel scourge / The good Brescians could not hold out / Against enemies to great slaughter / Because they were outnumbered / And every *schioppetiere* was lacking.'[88]

The reference to rain here draws attention to an important limitation of gunpowder weapons, often raised by their critics: they were vulnerable to the weather. As Sir Roger Ascham pointed out in a 1545 treatise on the virtues of archery, the Spanish had been defeated by Turkish archers at Castelnuovo when they 'had no use of their guns, by reason of the rain'.[89] The story of rain at Brescia was repeated in a later account of 1585,[90] and other narrations of the sack testify to the presence of Brescian arquebusiers, including those of the French writers Robert de Fleurange and Jacques de Mailles.[91] It is clear that already in 1512 both the French and local Italian troops had a good grasp of the use of firearms and their limitations.

The Sack of Rome, 1527

The expansion of firearm technology in Italy went on apace. In January 1514 (1513 old style), the officials of the Venetian Arsenal tested arquebuses with different mixtures of powder.[92] In 1517, the duke of Urbino hired *schioppettieri*; that same year many firearms were to be found in the stores in Treviso (part of the Venetian Empire).[93] Lists of *schioppettieri* drawn up for the duke of Ferrara in 1521 are dominated by Italian names, not only from the duke's own territories but from Milan, Bergamo and Bologna, among other locations.[94] That same year, the statesman and historian Francesco Guicciardini, then organizing financing and munitions supplies for a mixed army of Spanish, Italian, Swiss and German troops encamped outside Parma, lamented that the Spanish *schioppettieri* were using so much gunpowder that it was not possible to satisfy their demand.[95] Similar tactics to those used at Cerignola enabled the Spanish to prevail at Bicocca in 1522, while the Roman census of 1526 documents the presence in the city of an arquebusier named Guillelmo, and two *schioppettieri*, Francisco and Andrea.[96] The attachment of these individuals to those roles was clearly considered sufficient (by them or by the census-taker) to merit specific mention in the records. By the end of the decade, records of Italian troops serving in Apulia show that in many companies more than half the men were arquebusiers or handgunners.[97] Guns became increasingly common in policing, too. Benvenuto Cellini described in his autobiography an incident in Rome during 1529 in which his brother was shot by an arquebusier, subsequently dying of his wound. The shooter 'had once been a light cavalry soldier, and then he had joined the chief constable's corporals as an arquebusier'.[98] Cellini took revenge by stabbing him: at this early stage, when concealable guns remained few and far between, they were by no means always the weapon of choice.

More generally, Cellini's autobiography offers a good deal of insight into what was and was not possible when it came to guns in this early period. Cellini—writing well after the fact—cannot be regarded as a reliable narrator of his own life, but he offers a number of clues about guns and their limits. In 1527, troops loyal to the Holy Roman Emperor Charles V (ruler of both Spain and the

Habsburg territories in Germany and the Low Countries), sacked the city of Rome. Cellini recounted how he and his companions saw

> Bourbon's impressive army . . . exerting all its strength to force a way into the city. At the part of the wall we came to, many young men had already been killed by the attackers, the fighting was desperately bitter, and the place was covered by the thickest fog imaginable.[99]

This fog (from the context, no doubt partly a product of the gunfire) prevented them seeing precisely what was happening, but they each fired twice over the wall. Cellini pointed his gun 'towards the thickest and most closely packed part of the enemy, taking direct aim at someone I could see standing out from the rest' and found that 'the enemy had been thrown into the most extraordinary confusion, because one of our shots had killed the Constable of Bourbon'.[100] On the basis of later information, Cellini believed this was the man standing out from the crowd at whom he had been aiming, but he was not sufficiently confident amid the smoke to assert that he himself had been responsible for the fatal shot. Cellini went on to take charge of artillery in the Castel Sant'Angelo (where Pope Clement VII and his cardinals were besieged for several months), claiming impressive levels of accuracy, albeit with larger guns than the arquebus. As a practised shooter of wildfowl, Cellini was likely able to calculate trajectories at speed.

As Cellini's account of confusion amid the smoke suggests, a key change to the battlefield as firearms became more prominent was a growth in both smoke and noise. In a French account of the Sack of Brescia, Jacques de Mailles described his countrymen as making so much noise the arquebuses couldn't be heard, an image that relies on a shared understanding of the volume produced by a gun shot.[101] In a description of the 1525 Battle of Pavia, the Spanish historian Pedro Mexía likewise hinted at the physical sensation that this combination of sound could evoke: 'So great were the clamor and noise of the voices of the soldiers, of the clashes and blows they exchanged, and of the artillery and arquebuses, fifes and drums, that the air clattered and the ground seemed to tremble.'[102] On the whole, however, the sound of small arms fire receives less comment in the sources than that of artillery, which was frequently compared to thunder. Recounting the 1497 bombardment of Orvieto, cathedral canon Tommaso di Silvestro wrote in his *Diario*: 'First the fire is seen, then the smoke and immediately the thunder. The smoke lasts a long time.'[103] In Guicciardini's account of a 1510 speech by an ambassador of Vicenza, the envoy explained that it was 'the clash of weapons and the thunder of the cannon, noises to which our ears had never been accustomed, precipitated our surrender to the Venetians'.[104] Sixty years later, as the conflict drew to a weary end, the French poet Joachim du Bellay, working as a secretary in Rome, warily observed the shifting soundscape, transferring the metaphor of the storm elsewhere: 'You hear nothing but drums,' he wrote, 'and it seems a tempest.'[105] Drums were used to provide a marching beat, something that was particularly important for infantry moving in formation, including the new pike-and-shot units. An unusually positive account comes from Cellini, writing of his own shooting from the tower of the Castel Sant'Angelo during the Sack of Rome, who described how his usual artistic efforts were now directed into sounding the guns: 'My drawing, my wonderful studies, and the beautiful sound of my music: they were all in the sound of that artillery.'[106] I suspect, however, that this worked as much by shock value (a characteristic of Cellini's taboo-breaking style) and that the soundscape of gunfire was a less than lyrical experience for most.

Writing about guns

No one, to my knowledge, wrote a substantial text focused on guns in the early years of the sixteenth century, but discussion of firearms can be found in works of various genres. The destructive nature of the new weaponry was certainly apparent to those tasked with treating the wounds it caused—the surgeons—who took some decades to settle on the best approach. For a while, it was assumed that cauterization was best practice, and this was recommended in a chapter on the treatment of gunshot wounds included in Giovanni da Vigo's *The Practice of Surgery*. This was a widely used surgical manual, published in 1514, as handguns became increasingly common in warfare and indeed beyond. Giovanni, however, did not underline the importance of removing fragments of shot.

Earlier authors had been aware of this, but it was not until 1536 that Ambroise Paré, treating injuries during the Siege of Turin without access to the necessary oil for cauterization, found that outcomes were in fact better without it.[107]

In contrast to the medics, it does not appear that theorists of war in this period saw guns as posing any particular new problems. Francisco de Vitoria's foundational 1539 text *On the Law of War* makes no direct reference to small arms; Girolamo Garimberto, as we will see, is a little sniffy about the smallest handguns but accepts their presence; the baron de Fourquevaux writes extensively about their use in conflict in an entirely practical manner. Questions about the intrinsic morality of certain types of weapon have largely been the province of the twentieth century: it was only from the nineteenth that *jus in bello* (i.e., the law concerning the conduct of war) came to the fore in discussions of conflict. Earlier scholars focused principally on *jus ad bellum* (i.e., the moral justification for waging war in the first place): if a war was just, then all manner of activities in its pursuit were permissible, though they should be pursued with a sense of proportion.[108]

The rise of print, another of the new technologies of this period, enabled the circulation of military knowledge. These texts were just part of a wider literature on warfare, some of which dealt with firearms more explicitly. In 1546, Niccolò Tartaglia of Brescia published a compilation of advice in dialogue form on the use of guns and ammunition. *Quesiti et inventioni diverse* (*Various Questions and Inventions*) dealt in some detail with the practicalities of ballistics and recipes for gunpowder, its examples focusing on both military and hunting contexts; it had a revised and expanded second edition in 1554 and parts of the work were subsequently translated into English.[109] Tartaglia, who had been injured during the 1512 sack, explained that he was a theorist, not a practitioner; nevertheless, this was a thoroughly practical work.[110] Other such textbooks included in Italy Vanoccio Biringuccio's *De Pirotechnia* (*On Pyrotechnics*) (1540), on metallurgy, along with Battista della Valle's *Libro continente appertenentie ad capitanii* (*Book Containing Things Pertaining to Captains*) (1529) and Giovan Battista de' Zanchi's *Del modo di fortificar le città* (*On the Way to Fortify Cities*)

(1554), both of which addressed the question of fortifications. In Germany, literature of this type included Martin Merz's late fifteenth-century treatises on ballistics, Albrecht Dürer's 1527 book on the fortification of cities, castles and towns, and Fronsperger's *Vonn Geschütz und Fewerwerck* (*On Firearms and Pyrotechnics*) (1557).[111] Still another type of book focused on debates around the ethics of war and military techniques. This genre includes Fronsperger's *Geistliche Kriegsordnung* (*Righteous Rules of War*) (1565) and Roger Ascham's *Toxophilus*, a defence of archery (1545). Others still used print to make the case for peace, including, once again, Erasmus.[112]

Of the Italian texts on war, the most famous is undoubtedly Machiavelli's *Art of War*, published in 1521, which got an English translation before the end of the century.[113] Written in dialogue form, it makes limited references to firearms: in the voice of Fabrizio Colonna (a prominent condottiere), Machiavelli observes that the young men of a state should be drilled in the use of the arquebus, 'a new instrument . . . and a necessary one' alongside the bow and crossbow; he has Colonna suggest a 1:5 ratio of arquebusiers to other infantrymen, which is consistent with the wider trend for growth in proportion of shooters.[114] Machiavelli/Colonna also makes an important point about the potential for guns to create fear, arguing:

> I would want all the light cavalrymen to be crossbowmen with some arquebusiers among them who, although in the other managements of war are not very useful, are very useful in this: frightening peasants and removing them from a pass that had been guarded by them. For one arquebusier will make them more afraid than twenty other armed men.[115]

At the time Machiavelli was writing, guns were still poorly adapted for use on horseback, but the observation here is noteworthy because it echoes wider discussion of the gun's use in inculcating fear, an argument relevant to uses of firearms in crime and vendetta (to which we will come in part II) and frequently made in imperial and colonial contexts (discussed in chapter 9). None of the references to firearms in the *Art of War* is particularly novel or surprising: the text reflects the current situation at the time of writing.

The later years of war

Over the following decades, however, firearms became increasingly important in the conduct of war. By the end of the major land wars in 1530 (the year of Charles V's coronation as Holy Roman Emperor at Bologna), the Spanish had reinforced their power in northern Italy, with allies in Mantua, Florence and Milan forming a geographical barrier to any military attempt to revive the old French claim to Naples, now ruled by Spanish viceroys. An inventory of the Orsini fortress at Bracciano dated 2 January 1531 records the presence of fifty arquebuses and three bundles (*mazi*) of match-cord.[116] Small arms were used in naval warfare, too. A report of a fire at the Venetian Arsenal in 1534 noted that among the munitions destroyed were weapons to arm fifty galleys, including 'cuirasses [a type of upper body armour], lances, swords, *schioppi* and other similar arms'.[117]

As firearms grew in importance as a military technology, this necessitated an increase in supply. The Brescian documents (discussed further in chapters 2 and 3) show substantial expansion of orders in the 1530s and 1540s. The conflicts of these decades, and of the 1550s, were relatively small-scale: no longer did Italy see the large pitched battles that characterized the early stages of the wars and in which the pike-and-shot formations had proven so effective. Still, a 1547 report from Siena noted that that city was rapidly arming with a great quantity of 'arquebuses, pikes and morions'.[118] In the military context, guns were becoming the norm. The 1557 infantry muster rolls from Ferrara suggest that arquebusiers generally accounted for about a third of the infantry and confirm that by this point the personnel involved were primarily Italian.[119] In 1587, a total of fifty-five large arquebuses and *schioppetti* (in various states of repair) were to be found in the fortress at Castro (a Farnese feud north of Rome).[120] By the end of the century, some specialist militia units in Savoy included as many as 50 per cent arquebusiers and a further 25 per cent musketeers.[121] Indeed, by 1570 one military theorist was arguing that two-thirds shooters was the correct proportion.[122] While these were overwhelmingly infantry, by the middle of the century there were examples of cavalry units armed with arquebuses, following the example of the Reiter of the Holy Roman Empire.[123] (It was in the wars of the 1540s, Brugh notes, that 'the cultural impact of wheellocks was first felt in German lands . . . when mounted pistoleers became known as Black Riders [Schwarzreiter] because of their black battle vestments, which included holstered vests for transporting their loaded pistols'.[124]) Cavalry use of handguns may have emerged even earlier in eastern Europe: a letter to Pope Clement VII from Paolo Giovio on the Duchy of Moscow, first published in 1525, described its ruler Basil as having 'a band of *schioppettieri* on horseback' as well as 'many artillery pieces made by Italian masters'.[125] The proliferation of handguns did not gain whole-hearted approval: Girolamo Garimberto, whose treatise on the proper conduct of soldiers, *Il Capitano Generale*, was published in Venice in 1556, observed that 'many Italians, Frenchmen and Spaniards use an arquebus, and all the Germans carry two, or three small arquebuses, of which I wish I had not spoken, they being arms more of the street assassin than of the man of war'.[126]

Thus, from the fourteenth century onwards, gunpowder weapons were of growing importance in European conflict. To the early heavy artillery smaller arms were soon added, and by the middle of the fifteenth century handguns were in regular use on the battlefield, albeit in limited numbers. From the outbreak of the Italian Wars in 1494, firearms grew in importance and were particularly significant in victories at Cerignola in 1503 and Bicocca in 1522. As the conflict continued, numerous Italian powers established citizen militias and equipped them with firearms, a move accompanied by rhetoric around the defence of civic liberty (the militias are considered further in chapter 5). Despite this wide appreciation of the utility of guns in both offensive and defensive contexts, however, their use remained contentious. Francesco Guicciardini, who wrote a history of these Italian Wars, described artillery as 'diabolical rather than human', an image also used in relation to smaller weapons.[127] Even those involved directly in military command or weapon design, such as Blaise de Monluc and Leonardo da Vinci, could be ambivalent. Yet, despite the criticism, the military use of firearms grew until by the middle of the century around a third of infantry might be arquebusiers. Effective battlefield use, however, required reliable production and effective lines of supply, which in turn necessitated the development of an arms industry.

CHAPTER TWO

The Arms Industry of Brescia

NONE OF THE military developments discussed in chapter 1 would have been possible without a handgun industry. Private small arms production, however, added its own momentum to the firearm revolution. The people producing and distributing handguns acquired significant leverage with governments as demand for their products grew, while states were quickly faced with the question of how to sustain production capacity in times of peace. The simple answer was to allow exports to other states and private sales to merchants, but both options brought with them new challenges, especially in light of the high level of political and religious turbulence in sixteenth-century Europe, where, following the Reformation, conflict between Protestants and Catholics had added a new dimension to both civil and international wars. To explore these dynamics, this chapter takes as a case-study the small city of Brescia, which, thanks to the gunmakers of a nearby town, Gardone Val Trompia, became the single most important centre for the handgun industry in sixteenth-century Italy, a location at the heart of the firearm revolution.

Brescia lies about 180 kilometres west of Venice, in the far reaches of what was that city's mainland empire. It is, in fact, closer to Milan, and, as we saw in chapter 1, was the focus of conflict between the Venetians and the French in the first quarter of

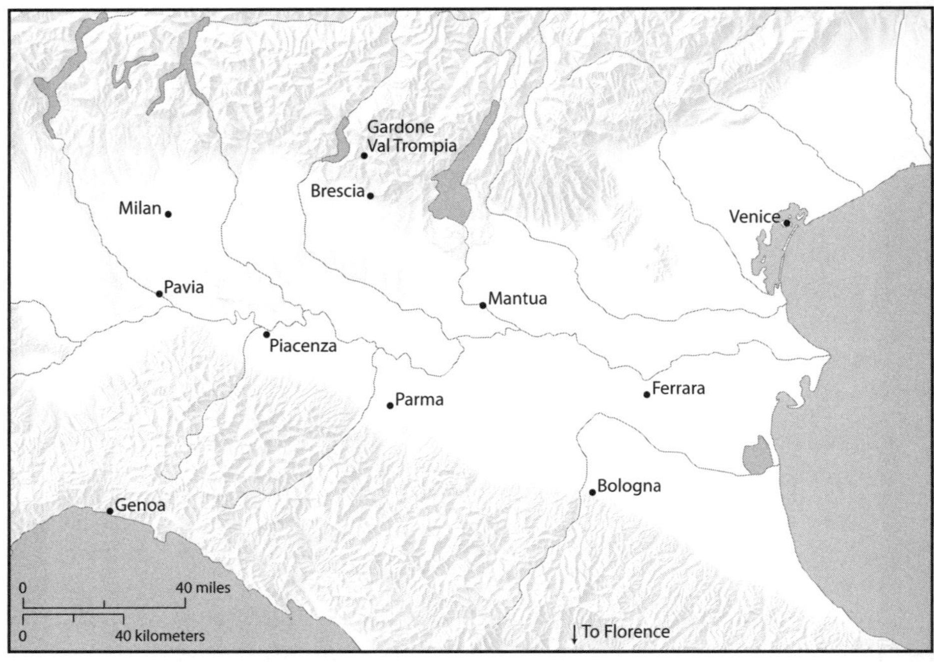

FIGURE 2.1 Map of Brescia.

the century. To its north are multiple valleys in the foothills of the Alps, including Valtrompia, where we find Gardone. Brescia was one of many production centres that grew up across Europe as rulers sought to secure their own supplies of small arms, complementing more established locations such as Malaga, Malines, Tours and Liège.[1] There has been continuous firearm production in Gardone for over five hundred years, in part thanks to some deliberate revival by the united Italian state in the later nineteenth century. In the early modern period, arms from Gardone were transported across Europe and quite probably beyond: the local case-study has global significance.

Gardone is also important because its story offers a counterpoint to a better-known institution of Venetian military provisioning: the Arsenal. The development of that public shipyard, dedicated to the production of galleys and organized in a manner that foreshadowed the modern factory, was notable.[2] Yet its reputation has overshadowed the very different trajectory of the firearms industry. Handgun production was certainly regulated by the state, through taxation and export licensing, and its fortunes depended to a substantial degree on government contracts, but fundamentally it was a private business, as we see in the 1526 order for 185 handgun barrels issued from the Venetian Arsenal to Bartolomeo Beretta.[3] While today military production is a matter for large corporations, in their early years handguns were produced by much smaller-scale private enterprise, a factor they have in common with the other prominent new technology of this period, printing. While printers benefited from state contracts, private investment was vital to their industry's development.[4] Guns, in fact, fitted very well into an already established world of military contracting, in which essential supplies for the conduct of war—both personnel and materiel—were obtained from private individuals and firms. This contracting approach also differentiated small arms from, say, cannon (which were typically produced centrally, sometimes even on the battlefield), and from the major fortification projects that constituted the defensive side of the artillery race. In terms of the production model, small arms had more in common with other individual weapons such as pikes, bows, armour and swords than with larger guns. This had wide implications for their management: arms control was only ever as strong as the state, and the Renaissance states were often rather weak, which gave arms producers considerable clout.

The first part of this chapter, therefore, sets out the processes of firearm production, explaining the environmental and political context in which a local industry emerged and the trajectory of its development. The second part turns to how the Venetian state managed some of the tensions with its arms producers, specifically through an export licensing system and in the context of the movements for religious reform that followed the rise of Protestantism. To do so, it draws primarily on correspondence between Venetian officials. The archival record for this early period is patchy: for Gardone itself only a single volume of notarial records survives, covering the years 1490–1545, with a bias towards the later decades. While references to charcoal production facilities in these documents hint at the significant local industries, sadly they do not provide the basis for a systematic reconstruction of the community's social and economic structures.[5] However, the documentation that does remain illustrates not only the leverage enjoyed by arms producers but also the growing importance of the industry, both for Venice itself and in its dealings with allies, where export licences, then as now, became a tool of diplomacy.

The establishment of an industry

Gardone Val Trompia is located in the hills about eighteen kilometres north of Brescia in the Italian Lakes area and lies 326 metres above sea level, although the mountains on either side of the valley rise significantly higher. Valtrompia, which had been settled at least since the Iron Age, had a natural environment well adapted to the production of handguns.[6] Iron came from mines beneath its mountains, while their slopes provided wood for charcoal production; the River Mella powered the watermills that hammered out sheets of metal to be rolled into barrels. Mining may have begun there as early as the sixth century; statutes dating from 1341 set out a legal framework under which mineral resources could be exploited, and these were updated in 1436 following the establishment of Venetian rule.[7] The local minerals included not only iron but also silver: the castle of San Giorgio, built in the early fourteenth century for the Oldofredi family and overlooking the road to Valtrompia, sat above a mine called the 'Bus del Quai' that can still be visited today.[8] Such fortifications are testimony to the area's long-standing importance and the care taken to defend it. Its visitors included Leonardo da Vinci, who during one trip, perhaps around 1510, sketched a map of Valtrompia, showing its small towns distributed along the Mella.[9] By the time of that visit, handgun production in Valtrompia had been going on for half a century.[10] In the decades that followed, the arms supply chain came to feature a complex series of microbusinesses: a description of 1609 outlines the roles of different masters, responsible for large and small furnaces, for the barrel's production, drilling, finishing and sanding.[11] Further specialists produced locks and stocks, and all of them in turn depended on the labour of others to supply firewood for smelting and routine supplies of food and drink.[12] By the time of Leonardo's visit, however, tensions between the artisans of the valley and their Venetian overlords were already apparent.

In order to understand how negotiations between Venice and its arms producers played out, a brief introduction to the political structures is required. Since 1426, Brescia and its environs had been part of a Venetian Empire that stretched from northern Italy to the eastern Mediterranean via the Dalmatian coast. Venice, however, was facing increasing challenges from an expansionist Ottoman Empire, to which it had lost significant Mediterranean territory, and from western European powers, who threatened its holdings on the mainland. Relations between Venice and Brescia were overseen by the Council of Ten (a magistracy responsible for security and public order), and in particular its *capi* (heads); on a day-to-day basis the city of Brescia and its surrounding countryside was governed by two rectors appointed by the Venetian state. These were the podestà, who dealt with civil matters, and the captain (*capitano*), responsible for military and police business. These officials ruled alongside a city council of local men who retained significant influence in policy-making and in the administration of justice.[13] Power relations were complicated by the continuing presence around Brescia of a local feudal nobility, and by the existence of village *comuni* or councils. By 1562, in Valtrompia alone there were seventeen *comuni* covering thirty-eight settlements and around eighteen thousand residents.[14] In emergencies, these valley *comuni* would meet in a General Council. If this sounds complex, it was, and the complexity also created many possibilities for local power-brokers to play off one group against another. For example, feudal noblemen might actively seek the support of village *comuni* in their own disputes, and that in turn might cause concern among the Venetian officials.[15] Each of these different layers of authority had its own priorities and interests in relation to the growing handgun industry.

These differing priorities become apparent in a letter of April 1505 from the Venetian officials in Brescia, in which they reported that 'some masters of *schioppetti*, arquebuses and shot' had left Gardone Val Trompia for Domodossola in the Duchy of Milan (close to the present-day border between Italy and Switzerland). One of the masters, however, had returned due to a lack of available charcoal in the new location. His responses to the rectors' questions about the masters' emigration worried them enough that they reported these events to the Council of Ten.[16] The timing of this letter is notable, coming just two years after the Spanish victory at Cerignola discussed in chapter 1. The advantages of firearms on the battlefield were becoming more apparent, and the Venetians were evidently keen to retain the services of the local artisans, an approach that mirrored their close guarding of another important technological process, glassmaking.[17] The following year, 1506, the rectors once again reported negotiations with Valtrompia producers, this time via a middleman named Jacomo de Philippin (Giacomo di Filippino), who was brokering a purchase of shot (*ballotte*).[18] A key turning-point came in 1509, when Gardone tradition has it that a man named Pietro Franzini developed a water-powered mill that enabled quicker and more effective production of arquebus barrels.[19] The General Council of the valley *comuni* also met that year to debate the provision of troops to Venice.[20] The dating here is notable, because in 1509 Venice was soundly defeated by the French at the Battle of Agnadello, temporarily losing much of its mainland territory. The Franzini story situates Gardone gunmaking as a technology that would enable Venice to turn around its fortunes. Again, this makes an important point about the balance of power: Venice needed the military personnel and expertise of the valleys. The arms makers had opportunities elsewhere, and this was by no means a one-way relationship. On the other hand, the industry was vital to Valtrompia because the valley was unsuitable for much agriculture and the residents depended on trading their products in order to obtain food supplies from the lowland farms around Brescia.[21] This likely explains why there was no clear distinction between the *comune* and the arms makers: a separate guild was formed only in the following century.[22]

Although after 1509 Venice largely refrained from involvement in the land wars, except to recoup its lost holdings, the development of a handgun industry facilitated defence of those territories it recaptured from the French, enhanced its firepower in conflicts with the Ottomans, and gave it leverage with other states through the licensing of exports. Here Venice benefited from the fact that while other parts of Italy did have iron mines, the variety of ore produced in the Brescian valleys had advantages over those found elsewhere, as the Sienese author Vanoccio Biringuccio noted in his

FIGURE 2.2 Herri met de Bles (Il Civetta), *Landscape with a Foundry*, c. 1540. Oil on wood, 88 × 115 cm. National Gallery, Prague, O 68. © National Gallery, Prague / Bridgeman Images.

1540 book *On Pyrotechnics*.[23] An impression of the infrastructure that supported arms production during this period may be gleaned from a painting of a foundry by Herri Met de Bles (Il Civetta) that shows iron-ore processing (Figure 2.2).[24] The Brescian production methods were also more developed than those of Biringuccio's hometown. Siena produced its own heavier artillery, but so far as it is possible to tell from the fragmentary sources, small arms production was relatively modest between the 1520s and 1540s. For larger consignments of guns, Siena turned to imports, including from the 1540s Brescian weapons.[25]

Frustratingly, the archival record does not allow any reconstruction of how gun production proceeded in the decade or so after the 1512 Sack of Brescia. The French do not appear to have had control of the valleys, and the valley dwellers played an important role in resisting the French invasion.[26] On one occasion in 1512, it was reported that 'all Valtrompia has taken up arms' in the name of the Venetians.[27] In 1516, guards were placed at the south end of Valtrompia to defend it.[28] However, the conflict also appears to have expanded scope for banditry (perhaps exacerbated by economic disruption), and in February 1513 the Venetian rectors embarked on a round-up of 'murderers, hitmen and thieves', threatening the valleys with the withdrawal of all their privileges if they failed to hand them over to the authorities.[29] This is an early example of the risks to social order that came with a highly armed population in a politically turbulent environment.

With the re-establishment of Venetian rule in 1516, however, business with the Brescian gunmakers got back to normal.[30] Documents from the later 1520s and '30s show how the rectors went about setting up contracts for the Venetian state.[31] For example, in 1533 the captain, Jacopo Correr, delegated contracting for arquebuses to the local quartermaster (the *soprastante di questi munitioni*), and he in turn secured an order via two men from Gardone Val Trompia, Zambonino di Marendelli and Jacomino del Chino, who were acting on behalf of the Gardone residents collectively (they were described as the *ostensori* or exhibitors of the current inhabitants).[32] When the heads of the Council of Ten queried an (unspecified) aspect of the transaction, Correr assured them that he had been told by more than one trustworthy person that these two were 'the best and most experienced masters' so far as arquebus production was concerned. He was sending Zambonino and Jacomino to Venice, and they were going willingly 'to defend their honour'.[33] Correr's reassurance as to their individual expertise and trustworthiness may have been a reputation-saving device on his own part, but it fits with wider cultures of doing business in the period, where contracting took place in a hierarchical and honour-bound social environment that shaped decision-making.[34] On another occasion, ahead of the Venetian-Ottoman War of 1537–40, we find the captain tasked with purchasing a range of weapons.[35] He shopped around first in Brescia, but after difficulties in securing a satisfactory price, sent instead to Valtrompia.[36] Brescia was not only a centre for handgun production but had become a hub for securing wider military supplies, as we will see further below and in chapter 3.

By 1562, Paolo Correr, the Venetian podestà in Brescia, estimated that Valtrompia had eight furnaces and forty forges, producing twenty-five thousand *schioppi* for export each year.[37] The demand for wood to produce charcoal and thereby smelt the iron required for gunmaking was so high that in the previous decade it had caused problematic price increases.[38] By 1572, Valtrompia had twenty-four furnaces and two hundred forges, was producing three hundred arquebuses a day (close on a hundred thousand a year) and, according to Domenico Priuli, who had just left the role of captain, was 'most famous' for that industry.[39] This was midway through the Fourth Venetian-Ottoman War (1570–73), a likely impetus for the surge in production. While the barrels came from Gardone, as Priuli explained, 'the stocks are then made in Brescia, Milan, and other places where the barrels are taken'; he made use of the export licensing system to limit the export of barrels in favour of completed guns, giving Brescian stock-makers a competitive advantage over rivals.[40] The Venetian rulers, however, had to tread carefully in their negotiations with the producers: this could be a challenging relationship to manage.

Valtrompia's privileges

The importance of the Valtrompia gunmakers to Venice is apparent from their success in extracting concessions from the ruling power. Venice needed the weapons they could supply, but in order to maintain that capacity between conflicts it had to permit exports.[41] During the 1540s, Milan, Spain, the Papal States and England all purchased Gardone weapons. In 1547, for example, the Papal States obtained a licence to purchase three thousand arquebuses, although, as we will see, this did not guarantee that supplies would be available.[42] The problem of retaining skilled labour in Gardone persisted as other states expanded their production and sought to secure expert services. One incentive available to the Venetian officials was to offer concessions related to tax and militia drill.[43] For example, on 13 May 1530, the doge (Venice's head of state, elected for life) ruled that the general requirement for men from the countryside to come into Brescia once a month for military drill could be waived for arquebusiers from Valtrompia and Val

Camonica (the valley to the west, above Lago d'Iseo) in favour of their drilling at a more convenient location.[44] Three years later, we see a similar decree, reiterating that the arquebusiers were permitted to drill only twice a year, and within their own valleys.[45] The issue evidently arose again in 1549, when the ducal registers record a reiteration of the exemption from the usual requirement to carry out ten musters a year.[46]

While the demands of the valley representatives were expressed in relation to their historic privileges, the valleys' military significance was surely a factor in the desire of the doge and Council of Ten to ensure that the privileges were observed.[47] The historic privileges dated back to 1426, when Valtrompia and the neighbouring valleys had come under Venetian rule and like other Brescian *comuni* had been conceded a certain level of autonomy, a trade-off with their new overlords.[48] Marin Sanudo, later a senator known for his diaries of Venetian political affairs, noted that the community of Brescia had 'many privileges from the Venetians on account of their merits'. For example, after successful resistance to a siege in 1438, Brescia had been granted complete exemption from payment of the *datia lancearum*, a tax normally used to fund the army.[49] Their role in resisting the French invasion of the 1510s may well also have been significant. Past military service was frequently invoked in relation to another type of civic privilege: the bearing of arms.[50] Language such as 'benemeriti': 'deserving' subjects, suggesting a desire to flatter the valley dwellers, may be found in the Venetian rectors' *relazioni* (reports given on their return from a posting). In 1554, Marino Cavalli described the Gardonese as 'most affectionate' to Venice; in 1547, Marc'Antonio da Mula had characterized Valtrompia and two other valleys as 'most devoted'.[51]

There was, however, a degree of tension between the Venetian officials in Brescia and their superiors over the handling of this relationship. In September 1531, for example, having already written on the subject once, the doge wrote a second time expressing his 'not mild displeasure' in response to complaints that in borderline cases the officials had been imposing their own, unilateral interpretation of the tax privileges, rather than consulting the doge.[52] The issue recurred in January 1533, when a further letter from the doge to the rectors emphasized the importance of observing all the privileges, 'so that they [the people from the valleys] should remain content, and not have reason to come and complain in our presence . . . to our displeasure and the malcontent of our most faithful subjects'.[53] It was vital for the Venetians to maintain good relations with the valley arms producers: the 1533 disputes arose at the same time as the rectors were trying to secure an order for 250 arquebuses.[54] This letter makes the important point that the Gardonese did not have to wait for the Brescian officials to agree to their demands: they could (and evidently did) also make representations directly to Venice. Letters from later the same year suggest the rectors became more cautious about consulting with the heads in Venice when such matters arose, but similar issues cropped up again in 1548 and 1549.[55]

The carrot of privileges for the Gardonese was accompanied by a stick, in the form of Venetian efforts to restrict migration of skilled masters. As the use of small arms in warfare expanded, other Italian states sought to establish their own weapons factories, although, as we have seen, they were sometimes constrained to import raw materials. A report of 1543 said the Sienese militia was short of both men and weapons: the soldiers had neither 'arquebuses nor morions nor other arms'.[56] Two years later, the Sienese authorities were purchasing 'arquebuses of Lucca', strongly suggesting that production had been established in that city.[57] In 1548, more than forty years after the initial concerns over emigration of Gardone masters, the Venetian authorities were still trying to prevent poaching of their artisans, refusing permission to a man named Battista Riccabello to take *diversi maestri* (various masters) to Florence, and threatening any who had already left with the confiscation of their goods if they failed to return.[58] A broader attempt was made in December 1556 to impose penalties (including exile) on subjects who took money from or entered the service of foreign princes.[59] This clampdown should be seen in the context of wider concerns about foreign attempts to recruit soldiers from the area, including by the marquis of Marignano in 1552 and the duke of Ferrara in 1557.[60] It underlines that guns were not produced in a vacuum but in a broader context of military contracting.

There were limits to Venetian power, however. Prominent masters like Battista del Chino certainly had offers from elsewhere and too heavy a hand in negotiations from the Venetians might have increased the temptation to accept them. In 1542, Cosimo de' Medici, duke of Florence, attempted to poach Battista; the duke succeeded in obtaining his services by 1551, when Battista was engaged to provide Florence with an annual supply of nine hundred arquebuses and one hundred muskets (muskets are a heavier and larger-calibre firearm).[61] Battista del Chino also supplied hundreds of firearms to Siena following that city's rebellion against imperial rule in 1552, and, as we will see in chapter 7, another member of the clan entered the service of the Farnese dukes of Parma and Piacenza.[62] By 1560, Cosimo had further succeeded in headhunting the Brescian master Nicodemo Magnano, who received a five-year contract to produce arquebuses and other ironwork for Cosimo at a factory in the Florentine subject town of Pistoia.[63] Later that decade, however, we find Florentine officials corresponding about the need to chivvy along Nicodemo di Marco (presumably the same person) to do the work commissioned from him on Cosimo's behalf, a hint that the power in this transaction lay on Nicodemo's side.[64] Brescian expertise was clearly attractive, not only on the Italian peninsula but for the production of weapons to be exported across Europe. The scope to secure offers elsewhere enhanced the producers' power in their negotiations with Venice.

The export licensing system

The desire of other Italian rulers to establish their own gun production facilities was undoubtedly boosted by the restrictions that the Venetian state placed on exports of firearms from Brescia. On the one hand, these restrictions enabled Venice to secure its own supply. On the other, they allowed Venice to offer diplomatic favours. This was not, it should be said, a watertight system: merchants could and did purchase small arms for onwards sale. It did, however, allow for some regulation. I am sceptical that the mercantile exception made for a complete free-for-all: merchants still had to navigate customs checks and secure transit licences, and, as we will see, suspicious exports were sometimes stopped.

Exports from Brescia beyond the bounds of the Venetian Empire were in progress by the early sixteenth century. A mid-century chronicle from the city of Modena records the 1508 arrival of Duke

Alfonso d'Este of Ferrara there, accompanied by some hundreds of infantry and cavalry, among them 'around 50 *schioppettieri*'. They had been equipped, so it was said, with arms obtained 'in Brescia for Duke Valentino'. That was a reference to Cesare Borgia, who had died in 1503; the consignment had subsequently been bought by his brother-in-law, Alfonso, who was known for his interest in military technology and portrayed by artists including Titian and Battista Dossi with one hand on a cannon (see chapter 8, figure 8.3).[65] In 1526, the marquis of the nearby city-state of Mantua was informed by the Brescian rectors that in order to export two hundred gun barrels he would have to seek licence from the Venetian authorities.[66] The spread of Brescian weapons across both the Italian peninsula and indeed the continent of Europe is clear. A 1539 inventory of the Rocca di Bracciano, a fortress in the Papal States north of Rome belonging to the Orsini barons, records the presence of nineteen Brescian arquebuses; these were equipped with stocks in the Spanish style as well as their powder-flasks, priming flasks, and moulds for shot (Figure 2.3).[67] Whether these stocks were made in Brescia or added later is an open question, but the local masters were evidently adaptable: a 1571 contract, to which we will come in chapter 3, included a specific request for a Brescian supplier to provide stocks made 'in the Spanish style'.[68] Examples in the Royal Armouries and from the 1545 wreck of the *Mary Rose* show the type of maker's mark that would have enabled identification of Gardone guns.[69] (Examples of these weapons are shown in chapter 3, figures 3.3–3.5.) Accounts for the years 1562–65 show that the Papal States were also importing arms from Brescia, including the large arquebuses that were shot from a stand (*archibugioni da posta*).[70] This was the range of demand that the export licensing system sought to accommodate.

Managing exports was particularly important because on more than one occasion there were problems of supply. The records show that it was common for states and princes to obtain permission for a large export, which would then be divided into smaller transactions undertaken by their contractors. The total number of guns supplied under these licences, however, was often lower than the total number authorized, suggesting there may have been shortages, or at least insufficient surplus to make the Venetian authorities confident

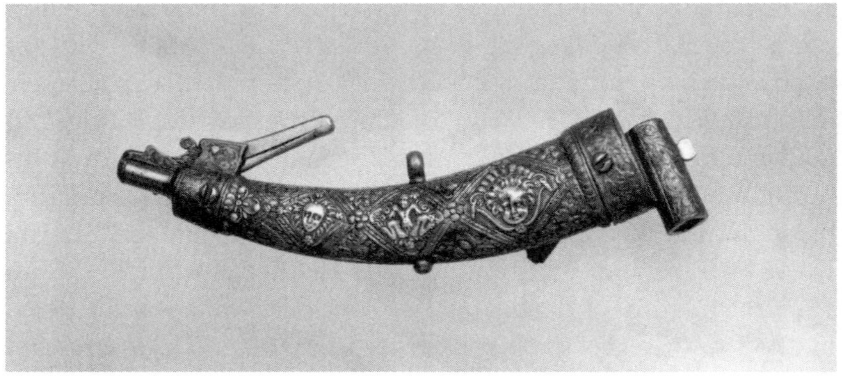

FIGURE 2.3 Unknown maker, priming-flask/spanner, France, c. 1570–80. Steel, gold, silver, 14.9 cm length. Gift of William H. Riggs, 1913. Metropolitan Museum of Art, 14.25.1523. Public domain.

that exporting was wise. While Venice had only limited involvement in the land wars of Italy after its defeat in 1509 at Agnadello, it had every reason to be cautious about French ambitions on the peninsula, not to mention the threat from the Ottomans to its territories in the Balkans and eastern Mediterranean. This was notably the case during the War of the League of Cognac, which between 1526 and 1530 pitted the majority of the north-central Italian powers, including Venice plus its former enemy France and (eventually) England, against the Habsburg forces of Spain and the Holy Roman Empire (including those of the Kingdom of Naples). On 20 March 1528, Bartolomeo Mazzoli, an agent of the duke of Milan's ambassador, was licensed to export 150 cannon (*columbrina*) balls, 200 arquebuses and 50 *some* of steel for the Duchy of Milan (a *soma* was roughly a mule-load).[71] However, while Mazzoli got the entire quantity of cannon balls on 21 April, he was restricted to the export of just twenty arquebuses and three *some* of steel; the following day he was permitted to export a further fifty arquebuses; still, this was barely more than a third of the number requested. On an unspecified later date export of a further twenty-eight *some* of steel was permitted, but that still left the steel export short of the total.[72] Even though Milan was an ally, Venice was evidently prioritizing defence of its own territory. We know this because on 8 April 1529, shortly before French defeat put an end to the League's ambitions,

Bartolomeo Parino of Valtrompia received authorization to export two hundred arquebuses to Venice. The doge, however, was clearly displeased when only 154 arrived, and on 14 April wrote to the rectors demanding they send as many as possible to Venice and forbidding exports elsewhere.[73] This was not only a problem during the major Italian land wars. In 1563, the captain of Brescia, Niccolò Gritti, wrote to report that having been requested to identify someone who might arrange supply of 1,200 large arquebuses (of the type that used a stand), he had secured an offer to deliver an initial 400.[74] That Gritti could not immediately promise to deliver the full quantity indicates the genuine process of negotiation between the state and its contractors: it was not only outsiders who had trouble obtaining all the weapons they would like.

A factor in the doge's concern during the 1520s was no doubt the domestic demand generated by Venice's decision to establish arquebus companies in its territories (on which more in chapter 5), but there were also significant worries about fraud.[75] Measures were taken to prevent dishonest claims for exemption from customs duty (*datio*), which was only granted to *comuni* when purchasing guns for their arquebusiers, and not when they were purchasing for onwards sale or any other purpose. The *comune* had to provide surety to that effect, and the rectors had to make a note on the licence what the buyer was permitted to export 'so that no fraud ensues'.[76] An individual exporter providing firearms to the Arsenal was also required to provide surety to the rectors, to be released only on confirmation that the entire quantity had arrived; the rectors were instructed not to allow themselves to be 'deceived and tricked'.[77] Bartolomeo Parino, whom we encountered above exporting two hundred guns to Venice, was permitted to do so only having provided 'sufficient security' (*sufficiente cautione*).[78] By requiring a financial guarantee, the state could mitigate against the possibility of exports going missing en route, but not without imposing an additional administrative burden.

Following the treaties of summer 1529 in which both the papacy and France made peace with the Emperor, arms controls were relaxed a little. In August 1529, the doge allowed infantry captains in Venetian service to stock up on arquebuses according to their contract (*condotta*, i.e., their general contract to supply military services) but without a specific permit, 'so that I don't have to write to you [the rectors] about this every day'. Still, he required the rectors to be 'diligent' that the arrangement was not abused, and to apply 'such limitations as seem prudent'.[79] Even assuming that there would now be a period of peace, there was a balance to be struck between control and efficiency. On the other hand, there were also new export possibilities, including to former adversaries. In July 1532, for example, the marchese del Guasto, an imperial commander, obtained permission to export from Brescia four thousand arquebuses, four thousand breastplates and five hundred muskets, which were transported in sixteen different consignments, the quantities of guns in each ranging from just seven to 464.[80] There were still limits, however. Later that year, when the viceroy of Sicily, a Habsburg appointee, was authorized to export 1,500 arquebuses, the rectors were told to advise him that he could not have any greater quantity.[81]

Securing adequate supplies was an ongoing challenge both for Venice itself and for foreign buyers. A letter of May 1554 from a man named Ercole Poeta, who was in Brescia to purchase armour on behalf of the duke of Ferrara, illustrates the problems. Poeta described a complaint from a Milanese man that the supply of morions he was meant to be obtaining for Genoa had been held up while the rectors sought confirmation from the Council of Ten in Venice that the export was acceptable.[82] A quite different gloss is placed on this story, however, by a contemporaneous report dispatched by the Brescian rectors to Venice, which accused the Milanese man of using the premises of a monastery adjacent to a city gate to smuggle sufficient armour out of the city to equip two thousand infantrymen. The merchant claimed this was a legitimate sale to a Genoese merchant dealing with Spain, but the rectors' suspicions were aroused by the unusual location of and secrecy around his storage facility.[83] In this case the export licensing system was evidently working to delay a suspicious transaction. Poeta further reported that the shops in Brescia were quite empty. This was due to the fact that in recent days the supplies had all been bought up by 'various merchants, who had taken them out of town'.[84] No surprise, then, that the authorities were proceeding with caution.

If the arms export controls made sense for Venice, they could be frustrating even for legitimate purchasers, as the correspondence of the papal nuncio to Venice demonstrates. (The popes at this time ruled a large area of central Italy with similar defensive requirements to any other state.) In 1541, the nuncio was Giorgio Andreassi, who had to manage a rather tense relationship between the papacy and the Venetian authorities over approaches to heretical books (of which more below). In April of that year, Andreassi wrote to Cardinal Alessandro Farnese describing his attempts to secure a licence to export arms from Brescia, and permission to transport metals via Piacenza.[85] When after considerable delay Andreassi finally obtained the necessary paperwork, which included a permit to export five hundred arquebuses, the cardinal asked him to thank the Venetians. Andreassi replied drily that if he thanked them after they had made him wait, 'then on whatever other occasion they will proceed even more slowly than they have done this time, especially with it not being their custom to do otherwise'. He did, however, make a show of gratitude.[86] Slow arms export approvals were one means for Venice to exert diplomatic pressure on an Italian power with which it was frequently in discord.

The challenge of religious reform

There was a wider context for this friction with the Papal States. Added to the existing push-pull of relations between Venice and its arms producers, the 1530s and 1540s were decades of religious ferment in Italy as the Catholic Church came under increasing pressure to respond to the challenge of Protestantism, the more radical varieties of which challenged not just the priesthood but wider social norms. We will come to the more general problems of public order and gun control in chapters 5 and 6, but the intersection of new religious thinking with access to weapons was a particular concern so far as it affected the arms-producing areas. In 1553, the Venetian rector Catterino Zen observed that in Gardone (among other places) 'everyone carries an arquebus and . . . they're not content with one, but even the women carry two, one in their hand and the other in their belt, both wheellocks, and they're a bad breed, untameable overbearing Lutherans'.[87] If back in 1512 the martial culture of the valleys had been an asset to Venice in its contest with the French, in the context of religious upheaval it risked becoming a threat to public order.

Zen's comment came in the context of a long history of religious tensions in Brescia as well as jurisdictional disputes between the papacy and Venice.[88] It did not help that in 1533 junior inquisition officials were reported to have committed 'great robberies and other disorders' in part of Val Camonica.[89] Such a heavy-handed approach may have encouraged the blasphemy that the Venetian government was keen to clamp down on in 1538.[90] The papacy and Venice, however, disagreed about precisely how strict to be. In 1541, the two sides clashed over the confiscation of 'several Lutheran books' from a Brescian bookseller, which the city rectors had asked to be returned. The nuncio, in contrast, was ordered by his superiors to ensure the books were burnt.[91] The following decade saw problems in Valtrompia, specifically around the detention of a man described by papal officials as 'a cursed heresiarch priest who with some others has turned Valtrompia upside down'.[92] This was Girolamo Allegretti, leader of a group of Anabaptists in Gardone.[93] (The slippage between Zen's Lutherans and these Anabaptists more likely represents the fluidity of definitions at the time than strict doctrinal differences.) When the nuncio complained to the Venetian College about the 'contagion of heresy' in Brescia, however, the College blamed the city bishop for failing to keep a suffragan (deputy) there. After assurances that one would be provided, the Venetians seem to have acted.[94] The nuncio emphasized that they were keen to secure 'a good and catholic preacher for that place Gardone in Bresciana, which is highly infected . . . to recover those souls of which many are contaminated'.[95] None of this, however, resolved the matter, and the appearance of some enthusiastic Dominican inquisitors in the area in January 1551 did not calm matters, prompting some 'malign heretics' to paste up flyers opposing them.[96] Conflicts over jurisdiction continued for several months.[97] Moreover, despite the nuncio's optimism that their efforts would 'exterminate the roots of this cursed sect', a decade and a half on there were continuing problems with heretics in Valtrompia, not to mention continuing tensions between the Venetian authorities and the Church, including over the religious attitudes of the rectors themselves.[98] On 28 June 1567, the nuncio in Venice was informed that once again in Gardone there was a 'congregation of heretics with great danger to the neighbouring peoples', and asked to persuade the Venetian authorities to have the Brescian rectors intervene, 'getting them to consider that those people who rebel against Lord God will much more readily rebel against their temporal masters'.[99] This surely played on the knowledge that Venice relied on the area for its arms production and could not risk that being jeopardized. The argument appears to have been persuasive, and the Venetians supported sufficiently firm intervention against the Gardone heretics to keep the pope happy.[100]

The sources here, of course, provide only one of multiple sides of the story. So far as I am aware, no direct testimony survives from the Gardone heretics, nor do the rectors' letters detail their response (particularly sensitive questions were often dealt with orally). In terms of gun production, the issues are far from straightforward. None of the nuncios' letters explicitly refers to Gardone as an arms-producing area, but, given that the papacy had previously purchased guns from Brescia, the links must have been known. One open question is the extent to which heretical beliefs were likely to have an impact on arms production. While there is evidence for Anabaptist use of firearms, notably during the Münster rebellion of the 1530s, later Anabaptists tended towards pacifism, refusing not only military service but also the payment of taxes to fund warfare.[101] The implication of Catterino Zen's report, however, is that in Gardone the religious radicals continued to carry guns. A second open question concerns the role of international trade in the transmission of new religious ideas. As we have seen, merchants travelled to Brescia to purchase firearms. Did they also bring with them news of religious reform? The wider archival documentation shows that local printers and booksellers had contacts in both Trent and Geneva.[102] Certainly, the Venetians were trading with England in the 1540s, a matter of some contention with the papacy. Their arms sales to England followed the break with Rome of 1533 (when Henry VIII had, however, maintained diplomatic relations with Venice), and came at a time when there was undoubted papal anxiety about Italian troops serving in England.[103] For Venice, political alliances and ensuring the arms industry remained sustainable evidently trumped some religious considerations. In general, Venice valued its reputation for relative liberty of thought, playing host to a range of religious freethinkers.[104] While the evidence for each specific case is fragile, the most plausible interpretation of the available sources is that the Venetians preferred not to police the Trumplini (residents of Valtrompia) too heavily, allowing some latitude when it came to religious dissent and arresting only when pushed, rather than imposing policing that would have upset the careful balance of autonomy in the valleys. Reliable arms supply from motivated producers was more important than who thought exactly what about God.

Conclusion

The case of Gardone arms production raises intriguing issues about the functioning of the Venetian Empire in managing and safeguarding its arms supply in the context first of conflict over the territory itself, and second over religion. The privileges granted to Valtrompia existed prior to the point at which gun production is recorded in the area, when new small arms were only just becoming known in Italy; they cannot be directly connected to the industry, although they may have recognized the significance of the mines in the area. However, the survival of these privileges into the sixteenth century may well have been connected to the growth of gun manufacture. The gunmakers of Valtrompia were clearly aware of their value to the Venetian state and worked collectively in negotiations with the Venetian representatives in Brescia to confirm their historic tax privileges and negotiate exemptions (for example) from militia drills, illustrating the power and agency that a subject area with key commodities and military production centres might exercise in relation to its overlords. Venice's control of arms production in Valtrompia, however, was challenged both by rival states seeking to establish their own factories and by religious change, where the imperial power had to tread carefully or risk alienating a key workforce in its own defence complex. However much the papal nuncios might have hoped the Venetians would impose religious order, the contractor state had other interests. In short, the firearm revolution and the consequent need for a ready supply of handguns shaped social relations in the Venetian Empire. The industry, moreover, offered opportunities to individuals at all stages of its supply chain, to whom we now turn.

CHAPTER THREE

Arms Dealers and Auxiliary Industries

THE NEW FIREARM INDUSTRY provided opportunities not only for manufacturers but for dealers, transport contractors and maintenance workers. This chapter explores the wider world of brokerage and auxiliary services that grew up to enable the use of firearms, which provided a vital link between gun producers and gun users, and which led to multiple individuals along a complex supply chain becoming engaged, one way or another, with guns. Just as the export of firearms was regulated through a licensing system, imperfect though it was, so their distribution was subject to state control through taxation and transit licensing. Once again, however, there were weaknesses in the system. Some officials took bribes; others were amenable to more subtle social pressure. While the modern ideas of black and grey arms markets were still distant, the problems created today by guns slipping out of official control would have been quite recognizable five hundred years ago, as would corresponding attempts by governments to prevent losses from the supply chain.

To explore how arms brokerage and distribution worked, this chapter focuses first on some remarkable surviving documents from the city of Brescia. These are the account books of Giovanni Battista Porcellaga, a local nobleman who facilitated arms purchases in the middle decades of the sixteenth century, and who thanks to the records' survival is one of Europe's best documented early arms brokers.[1] Porcellaga maintained good relations with local armourers, had connections in Gardone Val Trompia, and made an effort to cultivate members of prominent Venetian families. His case brings to life the functioning of the private contracts that were central to military operations in this period, illustrating the importance of social networks, gift-giving and patron-client relationships, and of the intersections between local politics and the international world of arms sales. Early modern rulers frequently contracted out elements of defence, buying in not only weapons but also the services of mercenaries. Military entrepreneurs, whether producers or brokers, and their employees or subcontractors, fulfilled this demand.[2] Yet beyond passing references to individuals involved in the supply of arms, most scholars have left to the margins of the literature the wider social relations that smoothed the way for arms purchasing.[3] Porcellaga's accounts reveal the people behind the rather dry terminology of the contractor state and show how the new products fitted easily into a social world where relationships were firmed up with the exchange of gifts and hospitality. The details of his brokerage appear in between purchases of fabric, horse tack and other household items. Indeed, these were probably not straightforwardly financial transactions: his reward came in the less tangible form of career opportunities for his son Scipione. While in today's popular culture the arms dealer is the archetypal Hollywood villain, I have found no evidence that such critiques were current in the sixteenth century.[4] Insofar as there was debate about the military contracting system, it centred on the advantages and disadvantages of mercenaries.[5]

Beyond brokers like Porcellaga, moreover, were many more individuals and organizations. The second half of the chapter turns to the afterlife of an arms deal, exploring how guns were transported and distributed, and considering the people who facilitated the movement of weapons, as well as the ancillary services that were essential to maintain arms and armour once they had been supplied. Those involved in the supply of firearms for military use in sixteenth-century Italy ranged from gunmakers to ropemakers, from bankers to customs officials, from city captains to leather-workers and scrap metal dealers. Through an exploration of the chain of supply and maintenance, the chapter offers further 'bottom-up' analysis of how the contractor state worked in practice. Its findings also have implications for our wider understanding of guns in society. As firearms became embedded into business networks, and as ancillary industries grew up to support them, more and more people beyond their direct producers gained an interest in their ongoing trade and circulation. That in turn underpinned the firearm revolution.

A broker and his network

As a broker of arms transactions, Giovanni Battista Porcellaga needed contacts with buyers, sellers and the licensing authorities. In securing them, it no doubt helped that he came from one of the older noble families of Brescia. He was born around 1492, and his son Scipione was the subject of a classicizing portrait now in the Museo del Castelvecchio, Verona (Figure 3.1).[6] The high social standing of the Porcellaga in Brescia is indicated by the fact that they had permission to bear arms (in practice, this meant wearing a sword) in both their home city and Venice; in 1551–52 Giovanni Battista served as podestà in the small town of Asola (like Brescia under Venetian rule).[7] The family history was not an entirely happy one, however. In 1538, a Zuan di Porcelagi [Porcellaga] was murdered by his *nipoti* (nephews or grandsons) for an inheritance.[8] The person responsible was exiled and does not seem to have been a close relative of Giovanni Battista. Still, the incident gives an impression of the volatile environment in which Giovanni Battista was facilitating arms transactions (we will return to both this and the privileges in chapter 6).

So far as arms purchases were concerned, Giovanni Battista's most important contacts were with the Orsini, a large Roman baronial family with multiple branches and a long history of service as condottieri.[9] Condottieri were military contractors, who took on contracts (*condotte*) to provide and lead troops. Those at the top end of the market like Camillo Orsini, whose purchases Giovanni Battista facilitated, might be responsible for the armies of an entire state (Figure 3.2). The Orsini were intermarried with the ruling families of both Florence and Urbino; Camillo's father Paolo, also a condottiere, had been in the service of Cesare Borgia until he

FIGURE 3.1 Alessandro Bonvicino detto Moretto, *Scipione Porcellaga in armature all'antica*, c. 1551. Oil on canvas. Museo di Castelvecchio, Verona. Photo: Rita Guglielmi via Alamy.

FIGURE 3.2 Dominicus Custos (printmaker), *Portrait of Camillo Orsini*, Augsburg, 1600–1604. Engraving on paper, 173×123 cm. Rijksmuseum, RP-P-1909-4414. Public domain.

was accused of conspiracy and put to death.[10] Camillo joined the papal service under Leo X (Giovanni de' Medici, r. 1513–21), and switched to Venetian patronage in 1522, following in the footsteps of Niccolò Orsini, count of Pitigliano (1442–1510), whose service to Venice had secured him a number of feudal properties in the Brescian *contado* (the rural area surrounding the city).[11] Giovanni Battista's first recorded transactions with the Orsini, however, are not directly with Camillo but with his secretary, Marco Antonio

Vittorino, for whom in 1541 he commissioned a gilded belt, probably a sword belt.[12]

While today such gift-giving might be seen as a conflict of interest, in the sixteenth century it was entirely conventional, analogous to present-day tipping cultures, and indeed gifts were often a vital supplement to the basic stipends paid to secretaries, courtiers and officials.[13] By 1545, Porcellaga was making presents directly to the Orsini family, giving two boar spears to Ottavio Orsini, lord of Monterotondo and a senior member of the clan; in 1548 he gave gilded swords in velvet sheaths to Camillo Orsini's sons Latino and Giovan.[14] These were to be presented to the pair by Porcellaga's son Scipione, an indication of the involvement not only of Orsini's wider family but of Porcellaga's in the social networks that surrounded arms transactions.[15] The personal presentation would have enhanced the gifts' symbolic value as well as giving Scipione a reason to meet directly with members of the Orsini clan.[16] After the assassination of Pope Paul III's son Pier Luigi Farnese in 1547 (to which we will come in chapter 7), Camillo briefly took over as governor of Parma before taking on another of Pier Luigi's former offices as Captain-General of the Papal Armies. Later, in 1548, Porcellaga commissioned further gifts for Camillo's relatives.[17] He also continued to cultivate relations with the wider Orsini clan: transactions involving Isabetta Orsini and Valerio Orsini (relating to linens and horse armour, respectively) appear in the accounts for late 1549.[18] While there were plenty of other reasons for a nobleman to maintain good relations with an influential family, not least because they were potential patrons for his own relatives, through these transactions Porcellaga consolidated one set of the connections that would be needed for his arms brokerage. He also kept other options open: beside the Orsini, Porcellaga was in contact with representatives of the duke of Ferrara, including Ercole Poeta, whom we met in chapter 2 and from whom in July 1553 Porcellaga purchased fabric.[19]

Turning to the supply side of arms provision, Porcellaga established connections through numerous small-scale purchases of arms from local artisans. In May 1541, for example, he ordered a sword and dagger from 'Mapheo spadaro' (Mapheo the swordsmith).[20] Later the same year, he ordered accessories and a blade from another swordsmith, Benedetto, who over the course of the next decade went on to supply the swords and small dagger given to Camillo Orsini's sons, as well as supplying and maintaining other edged weapons for the Porcellaga family.[21] Porcellaga, moreover, had contacts in the main firearms production centre of Gardone Val Trompia. The earliest explicit reference to a gun in this set of accounts is not to a large arms deal, but a small transaction involving two individual weapons: Ludovico di Tor di Gardone Val Trompia left an arquebus as a pawn to secure a left-handed *schioppo* but subsequently redeemed it.[22] By 1553, when Porcellaga was brokering a large contract for Camillo Orsini, he was in a position to tell Orsini that he had 'jollied along the arms bosses as much as I could' so that Orsini's agent could secure the weapons straight away.[23] He evidently had a wider familiarity with the elite social networks of the valleys.

Buyers and sellers were not enough, however: successful brokerage of exports also required contacts with the Venetian authorities who could authorize them. Porcellaga's role in Venetian service and local connections enabled him to smooth the way for Orsini in export licence negotiations: he did so in 1551, liaising with the captain of Brescia to obtain permission for the export of muskets.[24] Here Porcellaga was no doubt aided by his past cultivation of relationships with key members of the Venetian patriciate. In 1545, for example, he had purchased a little sword and armour decorated with silver and gilding for a nephew of Tommaso Contarini, and a further gilded stiletto to be presented to a nephew of Francesco Venier.[25] These were two very prominent patrician families of Venice, and while there were, of course, wider reasons for a Brescian nobleman to stay on good terms with leading Venetians, their gifts underline once again that arms purchases took place in a broader social context. This relationship with Venice, however, did not always win Porcellaga friends: in May 1547 he brought to the rectors a seditious verse opposing Venetian rule (and specifically mentioning the Porcellaga) that had been put up in the city.[26] On the other hand, it would have made him a good choice of broker. So far as I can establish, the Gambara family, whose sympathies for the French in the 1510s made for rather tense relations with the Venetian authorities, recorded in their account books primarily purchases of arms for their own use, with one modest exception when they assisted a Venetian buyer.[27] Even they, however, made an effort to cultivate the rectors: in 1551 they put on a dinner for the podestà's son and 'some gentlemen'.[28] The provision of hospitality in this period was regarded as a social virtue, and writers on diplomatic culture noted its more practical application as a means to gather news and make connections.[29] Indeed, such events, along with the relationships of these men to the Venetian patriciate, also raise questions about whether these middlemen might have had the influence to exert pressure on the rectors.

The arms transactions

Thanks to these three sets of connections, Giovanni Battista Porcellaga was well situated to facilitate the purchase of arms and armour in Brescia, as a close look at one series of transactions will demonstrate. First came the licences, secured by the papacy in 1547 and 1548 for the export of a variety of arms and armour, including three thousand arquebuses.[30] Immediately following the issue of the latter, Porcellaga's accounts show that he brokered a large order of arms, armour and other military supplies for Camillo Orsini.[31] This consisted of 275 corselets, 300 morions, 300 blades for pikes, 100 mattocks, 100 shovels and 50 pikes, along with a *soma* (roughly a mule-load) of iron plate to line a door, costing in total 4,485 lire. The inclusion of the mattocks and shovels is indicative of the importance of earthworks in contemporary warfare. Porcellaga then divided up the order between a number of smaller suppliers. Some of the armour was purchased directly from armourers, but a substantial consignment came via another middleman, a merchant named Lodovico Gandino who supplied 147 corselets and 55 morions as well as canvas and cord. Benedetto the swordsmith, who had already undertaken several commissions for Porcellaga, supplied 240 pike blades, illustrating that artisans of this period had capacity to switch between the production of high-end individual weapons and larger more basic military supply. Porcellaga facilitated the purchase of services, such as gilding and engraving of armour, as well as of goods. He was evidently able to broker a comprehensive range of military equipment, securing it both directly from producers and through other contractors like Gandino.

The following year, 1549, Porcellaga brokered an order of twenty-four muskets for Orsini, who in this case was represented by Francesco Borni.[32] Borni was a Brescian engineer then based in Parma where Orsini had been governor, and his involvement highlights the significance of technical experts in the contracting process. The guns were to be supplied by a Signor Zoane 'ditto Janino' del Pizzo, of Gardone Val Trompia, for a total price of 68 gold scudi. The accounts describe Zoane as 'His Illustrious Lordship', implying he was of noble rank (also confirmed by the fact he was titled Signor while a master craftsman would have been called Maestro). It is not clear whether Signor Zoane was related to the Pizzoni family, a house of mercantile origins that had been inscribed in the list of Brescian office-holders prior to 1488, but he was certainly someone of relatively high rank.[33] Nor was he the only valley nobleman to play such a role: the Negroboni family also provided arms and men to Venice over several decades from the early years of the sixteenth century.[34]

The wider contracting process was generally approached as follows: once the agents had arranged the purchase, the specifics of the weapons to be provided would be set out in a contract. Attention to detail and insistence on quality is evident, and in some cases might be checked in advance via a sample sent from supplier to purchaser: a Florentine inventory of arms from 1554–55 refers to two large arquebuses (*archibusi da mura*, which would be fired while rested on a wall) with walnut stocks 'as a sample of a larger number that is to be produced'.[35] Samples were likewise sent in relation to a Venetian purchase of arms and armour in 1534 and a Sienese purchase in 1546; in 1556 the Brescian rectors dispatched a sample of saltpetre to Venice and awaited confirmation on whether they should proceed with the order.[36] In the case of Orsini's order, the specifications for the guns were as follows: they were to be the colour of iron, clean and in laudable form; forty-one ounces long (there were twelve ounces in a *braccio*, so these were guns of two metres or more); taking shot of three ounces; and weighing sixty *liri* each (this would imply they were larger muskets and weighing around eighteen kilograms).[37] Half were to be supplied with stocks in the Spanish style, and half with stocks in the Italian style; the order was to be accompanied by twelve flasks, twelve priming

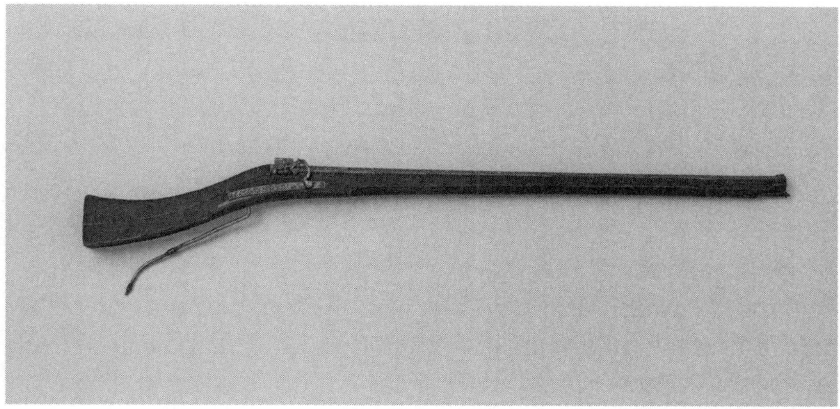

FIGURE 3.3 Unknown maker, matchlock rifle, c. 1545, Spain. Iron, gold inlay, wood, velvet, gold thread. 124 cm length. KHM, Hofjagd- und Rüstkammer, D 57. © KHM-Museumsverband.

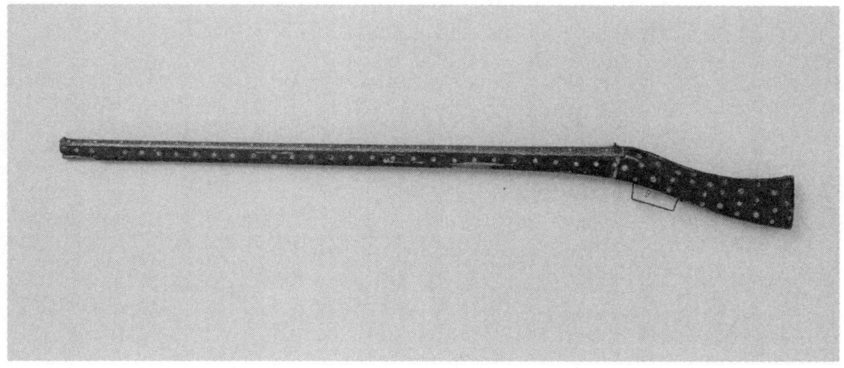

FIGURE 3.4 Unknown maker, matchlock rifle, c. 1545, Spain. Iron, gold inlay, wood, velvet, gold thread, brass nails in rosette shape. 135 cm length. KHM, Hofjagd- und Rüstkammer, D 156. © KHM-Museumsverband.

flasks, twelve moulds for shot and twelve *cavaballe* (a device for cleaning the barrel), along with a ramrod for each weapon. This level of detail was standard in contracts, as examples from 1546 and 1571 also show.[38] An indication of the Spanish style can be seen in figures 3.3 and 3.4, and the somewhat straighter Italian stock in the Royal Armouries Gardone weapon from the 1540s (Figure 3.5). Evidently, the Brescian producers had adapted to provide the preferred foreign style as well as the local one. It is also notable that the

FIGURE 3.5 Unknown maker, matchlock military musket, c. 1540, Gardone Val Trompia, Italy. Iron, walnut wood. 130 cm length. Royal Armouries, XII.5315. © Royal Armouries.

flasks, moulds and *cavaballe* were supplied in half numbers: this may indicate that these relatively heavy guns were to be used from a static position in which sharing accessories was a viable option. To secure the order, Borni paid a deposit of 10 gold scudi from Orsini's funds, which was received by an agent of Signor Zoane named Messer Francesco (a title indicating middling rank). Porcellaga's role was to facilitate the transaction: he did not directly take a cut.

The role of the export broker was all the more important in moments like the early 1550s when there was high demand for supplies. A follow-up order of August 1551, again for Camillo Orsini, again supplied by Signor Zoane del Pizzo and again for muskets, shows a rise in price to three ducats each from the previous 2.83 gold scudi.[39] (Scudi were generally of slightly lower value than ducats.) One master, Josepho di Paradisi, was able to command a price of 1,752 lire for a consignment of seventy arquebuses with their accessories (approximately 3.5 ducats each).[40] That supply was constrained around this time is confirmed by the fact that the rectors recorded exports of only 1,689 arquebuses out of a total of three thousand licensed for dispatch to the Papal States in 1547.[41] Just as Venice itself struggled to secure adequate supplies, so did international purchasers. Finance for a proportion of Orsini's purchases had been supplied by a Florentine merchant, Antonio Ubertini, who received his payment only in December 1552.[42] That raises many questions about cash flow and credit which unfortunately cannot be answered with the current sources. In light of wider cultures of credit in this period, however, which were highly dependent on social connections, it is very possible that part of the reason for Porcellaga's personal involvement was to act as an effective guarantor for the eventual payment, as well as to help secure supplies in a tight market.

What did Porcellaga get out of this activity? He did not engage directly in trade: to do so would have disqualified his family from participation in the Brescian council.[43] The criteria for council membership became increasingly aristocratic in the first half of the sixteenth century so that by 1546 even a grandfather's participation in trade or industry excluded a man from membership.[44] His family could, however, receive benefits, such as the loan of 427 gold scudi made in 1549 by Camillo Orsini to Porcellaga's son Scipione, who was raising five hundred light cavalry.[45] Indeed, it is possible that Scipione joined, or was hoping to join, Camillo's service; he certainly entered military service at a later date, and died in Cyprus during the Fourth Venetian-Ottoman War of 1570–73.[46] In terms of sixteenth-century European norms, there is nothing surprising here: this is the typical behaviour of provincial nobility trying to secure the favour of key members of an internationally powerful family. Giovanni Battista Porcellaga himself may have been granted the title of Lateran Count, an honour granted by the pope and sometimes attached to papal office-holding (which might in turn lead to financial benefit).[47] Porcellaga's choice of gifts for the men he hoped to influence is also notable. All sorts of presents were available, from food to textiles to art, but Porcellaga repeatedly made gifts of weapons. He did not give guns; he opted for decorative edged weapons that might be worn or (in the case of the boar spears) used by the recipient for hunting. We will see in chapter 4 that in some cases guns were given as gifts in the 1530s, but Brescian manufacturers were not yet routinely producing luxury guns, and perhaps Porcellaga preferred to play it safe given the continuing cultural ambivalence about the new weapons. Together, the evidence points to a cautious but consistent integration of the modern technology of guns into this pre-existing social world.

Transport and distribution

Once Porcellaga's buyers had secured their export licences, they would also need permission for transit of their arms consignments through third-party states, and that necessitated further diplomacy. A 1571 contract for firearms between the Papal States and a Brescian supplier stipulated that the guns were to be delivered to Ancona by the makers, but provided for the possibility that the rulers of Mantua or Ferrara might hold up the consignments.[48] Examples of transit licences may be found in the Mantuan archives: on 31 January 1543, for example, a Captain Belantonio Corso was granted an exemption from customs duty in respect of a 'certain quantity of armour' to be brought from Brescia for Pier Luigi Farnese; references also survive in Florence, where in 1552 the Sienese authorities requested a similar transit permit for Brescian arquebuses.[49] There were, however, ways around the licensing system. In 1531, a warrant was issued for the arrest of two former Brescian constables (one responsible for the gate of San Nazaro and the other for the gate of San Giovanni), who along with a local shop-owner and the wife or son of one of the constables, had been allowing a variety of contraband goods, including not only guns but also wheat, to be smuggled out of the city.[50] Constables, captains and indeed local residents more generally were another significant group of actors in the course of any arms sale, and their trustworthiness or otherwise was a key factor in the state's ability to regulate arms exports. The authorities faced a tricky challenge in balancing the benefits of efficient distribution of arms with the costs of control and monitoring.

Once authorized, the transport itself required further contracting. An account book for the papal fortresses, dated 1541–42, illustrates a variety of payments for the transport of goods, for example to 'Morgante porter for the transport of the said arquebuses from the Ripa customshouse to the barge', incidentally revealing the involvement of customs officials in the process.[51] It also shows the expenses incurred by the book-keeper and his party (including two horses and two servants) for transporting munitions to Paliano (about 60 km east-south-east of Rome). Payments might be made to muleteers, for carts, and for other unspecified transport as well as for paperwork such as a licence.[52] The mules would have required fodder en route, a further aspect of logistics to be considered, and not always a straightforward one given the devastation wrought on Italian lands in the course of extended warfare. While the 1571 contract specified that transport to Ancona was the responsibility of the producer, other documentation from Rome suggests this was not invariably the case.

Considerable administrative effort went into managing the transport process. Guns were valuable, and a threat to public order in the wrong hands. A bill from 1575 for 591 boxes of various arms dispatched from Brescia to Ferrara, including 9,300 arquebuses, shows that each box was numbered.[53] These inventory numbers were subsequently used for reference during the transport process, with notes taken of who had custody of which numbered boxes.[54] This was not a system unique to Brescia: inventory numbering was used by the Ferrarese authorities too.[55] Moving a 1575 consignment from Pesaro to Ancona alone involved an official at each end of the chain, and payments to a number of different bargemen, some of whom made more than one trip. A man named Giovanni Matteo Bruciaferro took forty-eight boxes of arms and armour on his barge on 4 May 1575, and payment is documented both to the bargeman and to porters for unloading and carrying. Over the coming week, several different bargemen took consignments ranging from twelve to a hundred boxes; however, later in the year a single bargeman named Alessandro Ferrarese became the principal contractor. There are different ways to interpret this change of pattern: it may be that a large initial consignment was divided between whoever could be contracted at short notice, but that subsequently one barge operator became the preferred provider; on the other hand, it is possible that in the autumn one operator took responsibility for the contract as a whole but then subcontracted sections. Once again, this stage of the arms supply process points to the number of people involved: not only those at either end of the supply chain but also the bargemen and porters, plus whoever made the boxes in the first place and the notary who eventually audited the accounts. For weapons to be effective, the logistics were vital.

The next step in the chain was distribution. In the Papal States local *massari* (town captains) were responsible for purchase of weapons on behalf of their communities and then their issue to the militia. In 1549–50, each *comune* received (and was required to pay for) arquebuses, pikes and halberds. Toscanella got fifteen arquebuses, as did Corneto; Civitavecchia on the coast got twenty, as did Tolfa; Bolsena thirteen, San Lorenzo fifteen again, Acquapendente twenty-five, Orte fourteen, and so on.[56] Similar arrangements were made for the distribution of firearms to the arquebus companies in the Veneto (on which more in chapter 5), while lists produced in Florence and Ferrara during times of siege (or feared siege) show the process of monitoring who received firearms and other weapons from central supplies (or already had their own).[57] The prospect that supplies might disappear or be removed from stores weighed on the minds of rulers. In May 1531, following the smuggling incident in Brescia, the doge of Venice ordered that inventory be taken of supplies in all the castles and fortresses of the Brescian territory, recording the quality, quantity, number, weight and measure of everything present 'so that no one can commit fraud in any part of the aforesaid stores'.[58] While the most explicit concern in that case was about grain, missing guns were certainly a problem. A 1558 inventory of the papal fortress at Perugia included a list of 'everything that is missing from the armoury'. Among the stray objects were six arquebuses. In some instances, the record-keeper noted that the missing items had been given out, but in other cases he appears to have been ignorant of the stock's fate.[59] These inventories were typically made at the point of handover between officials, for whom the list of missing items served to clarify responsibility for losses.[60] While careful record-keeping could not prevent the loss of firearms, it is evidence for the level of attention given to monitoring weapons supply and distribution.

Ancillary services

A wide variety of ancillary roles also developed to support the growing firearms industry. Guns came with accessories: armour adapted for arquebuses (made by armourers), holsters (by cloth- and leather-workers), ramrods and moulds for producing shot (smiths and metalworkers). The records refer to ropemakers (Rome), maintainers/repairers (*acconciatori*, Florence), scrap metal merchants (Rome), recyclers of old guns (Bologna), and flask-makers (Brescia). Morions were also produced in Brescia, as we saw in the Ercole Poeta letter mentioned in chapter 2. There were gunpowder factories in Venice, Padua and Brescia, where in January 1543 the rectors were liaising with saltpetre suppliers.[61]

However, skills in gunpowder production were also more widely diffused.[62] In September 1526, the heads of the Venetian Council of Ten issued a warning to one Francesco de Calabria (Calabria was part of the Kingdom of Naples) not to make or keep any kind of gunpowder or guns in his house on pains of losing the property.[63] An inventory of the Orsini fortress at Bracciano, taken in 1543, shows the presence of equipment for making gunpowder, including 'a large wheel with its iron handles, pestle and mortar' and 'a bit of coarse saltpetre in a barrel'.[64] Moreover, when the following year Francesca Sforza Orsini of Bracciano appointed a castellan, she placed a range of conditions on the appointment, including that among the eight infantrymen he was to hire there should be a bombardier who knew how to look after artillery and make gunpowder.[65] As we will see in chapter 9, once they secured the necessary ingredients, European soldiers in the Americas were also able to manufacture powder.

Export licences issued in Brescia frequently include alongside guns reference to both powder-flasks and the smaller priming-flasks (used to measure the appropriate dose of powder for the weapon) as well as the helmets (*morioni* and *celate*) typically worn by arquebusiers. When the papacy purchased arquebuses in 1575, they also accounted for both sizes of flask.[66] Matchlock arquebuses, moreover, required supplies of match-cord, which was sometimes bought from people specifically identified as ropemakers such as Federico, based in the Borgo (the district of the Vatican), who supplied 163 pounds of hemp for the fortress of Paliano in 1541–42.[67] On other occasions, however, hemp was purchased from merchants or indeed from other states. For example, in 1538 during preparations for a naval campaign against the Ottomans, the papacy purchased hemp from two prominent Venetian patricians: Paulo Giustinian, and Alessandro Contarini.[68] There may have been political considerations in this choice of suppliers, but it may also have reflected the convenience of working with men local to the port from which the armada would depart. Besides match-cord, the records detail a range of purchases of ironwork in 1541 from two Jewish *ferrari* in Rome, Abramo and Iosepho, including mattocks and pickaxes; a third Jewish supplier, Salamon, provided a large second-hand cauldron for munitions.[69]

While these suppliers capitalized on the growing use of firearms on the battlefield by offering complementary products, others looked to the need for maintenance services. Guns could easily become unusable if they were not regularly serviced. Of the three iron *schioppetti* identified in a 1522 inventory of the Rocca di Palo (then in the charge of Felice della Rovere Orsini, daughter of the late Pope Julius II), one was broken.[70] The twenty-seven large arquebuses listed in the 1543 inventory of the Rocca di Bracciano were old.[71] In a 1549 account of artillery in the Rocca di Carpi, there were thirty-eight large arquebuses 'unhappy, in poor order' and only seven 'good and well in order'; another undated note in the same file recorded the state of arquebuses in various fortresses, noting that water damage had 'ruined' all the arquebuses in Modena.[72] Such problems could be avoided to some extent with preparation and planning. In 1529, the Nine Conservators of Ordnance and Militia in Florence paid for powder and shot but also for 'fixing up' arquebuses.[73] Soldiers serving in the militia of the Duchy of Castro were, according to the 1568 statutes, required to ensure their weapons were 'in good order'.[74] In his essay on the military revolution, Michael Roberts suggested that firearms were 'forcing the soldier to be a primitive technician', but the existence of private maintenance contractors a century prior to his main case-study prompts questions about how much of this work was in fact expected from an individual soldier and how much might be contracted out.[75] A 1548 decision by the Venetian Council of Ten and a 1559 decree on arms-bearing in Bologna both acknowledge the existence of repair services,[76] while further evidence for their accessibility comes from the travel accounts of Cesare, a servant of the Gambara family, who in 1575 had an arquebus fixed in Ravenna.[77] Court records of 1577 refer to travellers stopping to have their guns serviced in Castel San Pietro, on the road between Bologna and Imola.[78] The similarity between wheel-lock and clock mechanisms meant that clockmakers, too, had opportunities to diversify into gun maintenance.

A document in the Orsini family archives, listing the contents of a workshop that appears to be involved in gun repairs, helps explain the need for specialist services.[79] It is described as an inventory of the 'bottica di Tartaglia' (workshop of Tartaglia), and it is tempting to associate the name with that of Niccolò Tartaglia, known among wider mathematical work for his calculations of artillery trajectories, and whom we met in chapter 1; however, as he himself described his work as entirely theoretical, that seems unlikely.[80] The document is undated, but the archive cataloguers date it to the first half of the sixteenth century, a conclusion with which I agree. It lists the following items:

> In primis vinti cinq[ue] archebusci acco[n]cj | Twenty-five fixed up arquebuses
> Item tre archibusci guastj | three broken arquebuses
> Item uno archebuscio quale d[e]ue haver[e] hauto LucaJoannj | one arquebus that LucaJoanni should have had
> Item tre martellj et una maza | three hammers and a mallet
> Item sei lime | six files/rasps
> Item uno raschiatoro | one scraper
> Item uno scarpello | one chisel
> Item diecenove tagliolj | 19 taglioli (a tool used by the metalworker to cut hot metal)
> Item una vite | one screw
> Item una anchudinetta piccola | one little anvil
> Item tre para d[i] tenaglie | three pairs of tongs
> Item due spine | two plugs
> Item quatro trapani no[n] finitj | four unfinished drills
> Item uno trapano finito | one finished drill
> Item uno trapanetto piccolo arota | one small drill with a wheel
> Item duo sfrecatorj | two scrapers.

While these do not appear to be complex tools, they are certainly greater in number than could feasibly be carried by an individual, and it seems likely that it was simpler to drop off a gun for maintenance at a local shop than for every gun owner to acquire skills in repair. Gun ownership did not only create a gun *manufacturing* industry but also generated new opportunities in ancillary services.

A reference to cleaning of halberds in the Rome records includes a rare mention of a woman involved in weapons maintenance: on 9 August 1549 two ducats were paid for this purpose to 'Madonna Elizabetta, the wife of Maestro Angiolo, a swordsmith'.[81] It seems equally plausible that cleaning of guns might have been subcontracted, and indeed that many of the payments to men found in the records will also incorporate work done by women. Certainly, in the eighteenth century the Gardone gunmakers employed female apprentices to burnish the finished weapons.[82] Whether there was modification of arquebuses beyond the basics of maintenance is another outstanding question, but this certainly happened in the civilian context (including to facilitate concealment of weapons), and it is quite possible that soldiers also attempted to improve their own kit.[83] In short, the use of firearms generated demand for a wide variety of goods and services, some specific to guns (repairs and modifications) and some general commodities which had other civilian uses (provision of hemp for processing into match-cord). The involvement of ropemakers, ironmongers and scrap metal dealers in the broader arms supply process raises important questions about producers' choices. Did they make a conscious decision to turn towards military supply in the hope of profit? Were they obliged—officially or in practice—to supply the state? The sources at hand do not permit a clear answer: rather they highlight the shifting pattern of demand during wartime and the need for states to find supplies to match.

Conclusion

Given the complexity of securing both supply of sufficient arms and the appropriate permissions, it is not surprising that we find members of the local nobility engaged in facilitating some of the larger purchases. Giovanni Battista Porcellaga had connections in Gardone sufficient to secure contracts, including via local lords (not just masters) and ongoing relationships with arms suppliers. The intersection of the contractor state with a gift economy in which presents eased social relationships and in which broader social networks were deployed to facilitate transactions is evident here. Guns may have been modern, and there were evidently some traders who purchased arms for resale in a quite straightforwardly commercial way, but personal relationships—indeed, given the status of Signor Zoane, perhaps feudal ones too—also mattered. Even as this very modern industry developed, it did so in a context of social relations characterized by much older structures, in which personal and collective credit and reputation shaped relationships, as did perceptions of loyalty to the imperial power. The state's ability to purchase effectively depended on local patrons and connections, and at times profiteering merchants seem to have taken advantage of state weakness. The private firms involved in arms production used their leverage to extract and maintain concessions and privileges from the state. Along the supply chain for firearms, there are places (as in the monitoring of transport and arms distribution) where the state seems to have been relatively strong. Elsewhere, however, it might be circumvented by corrupt officials or enterprising merchants. In short, what emerges in the military context is a picture of patchy and uneven development of state control of firearms supply, combined with a substantial element of private contracting that functioned on an international scale.

Moreover, the picture of handgun supply and distribution that emerges from the archive sources is one of complexity. Even before we reach the end users of these weapons, it is already clear to see that the firearm revolution engaged individuals from across society in numerous transactions: in purchasing raw materials for gun mills, in adding stocks and locks to gun barrels, in obtaining suitable accessories. Besides the conduct of the arms deal itself, which could involve a broker, technical experts, a purchaser, his staff and witnesses to the transaction, transport workers were required to assist with distribution, and officials to manage the final distribution to militia or armies; clerks and auditors ensured this purchase was monitored along the way. Ropemakers, gunpowder suppliers, holster-makers and scrap metal merchants contributed supplies; maintenance staff assisted with repairs. It would not be exceptional to find twenty-odd different steps involved in getting a gun from raw materials to end user. Many of these steps are found in other supply chains of the period, but the specialist nature of gun barrel production combined with the need to safeguard supplies in the interests of defence and public order made the management of this one an unusually charged matter. On the production side, the role of the state here was primarily one of manager and regulator of private contracting, but this required a high degree of bureaucracy to be effective. It is here as much as in the tax-raising process that we might look for a 'military state'. The state's interactions with its citizens and subjects were not only about extracting money, nor issuing contracts, but about creating and monitoring a system in which multiple actors could each play their part in the state's defence.

While the production of handguns was a private business, the risks associated with this technology favoured the development of a more active state. That was one impact of the firearm revolution.

The supply of arms was a business that operated in parallel with an elite culture that remained suspicious of the new weapons. That did not, however, mean that the men involved in their brokerage were in any way a separate class. Wider purchases in the Porcellaga account books show the family's broad humanist culture: they purchased copies of Virgil's poetry published by the company of Aldus Manutius for Giovanni Battista's sons Marcantonio and Theseo.[84] Scipione Porcellaga would find himself the subject of a classicizing portrait. Yet guns were not an easy fit with the revival of ancient culture that characterized the Renaissance. We turn now to see how Italian elites sought to square that circle.

PART II

Beyond Warfare

CHAPTER FOUR

Court Culture and Luxury Weapons

THE GIFT-GIVING THAT helped develop and consolidate Giovanni Battista Porcellaga's social network also operated on a much grander scale. It was one means by which firearms—and new types of firearms—were introduced to the court societies of northern Italy. In 1538, Francesco Maria della Rovere, duke of Urbino, wrote to the duke of another small city-state, Mantua: 'I am very grateful to Your Illustrious Lordship for your reply to me regarding the arquebus that should be coming from Germany, and for the barrel you have sent me which shoots without a match and is very dear to me.'[1] Such appreciation for novel firearms like this self-firing gun barrel (most likely a wheellock weapon) runs counter to the dominant image of guns in the European literature of this period, which was generally hostile: in text guns were the Devil's work, the coward's weapon.[2] The duke, however, was not alone in his enthusiasm. In his memoirs, Benvenuto Cellini described his pleasure in firearms, in the context of an encounter in the same decade with Duke Alessandro de' Medici of Florence:

> While we were thus talking, his Excellency was in his wardrobe, looking at a remarkable little gun that had been sent him out of Germany. When he noticed that I too paid particular attention to this pretty instrument, he put it in my hands, saying that he knew how much pleasure I took in such things, and adding that I might choose for earnest of his promises

> an arquebus to my own liking from the armoury, excepting only this one piece; he was well aware that I should find things of greater beauty, and not less excellent there.[3]

It would be tempting to regard this as Cellini's typical exaggeration, or as inflected by hindsight from the later vantage point of 1558–63 when he was writing,[4] except that whoever made the inventory for Alessandro's successor also described certain guns as 'bellissimi' (most beautiful).[5] Alessandro had begun acquiring guns at the age of around thirteen; in 1553, at the age of twelve, Paolo Giordano Orsini was equipped with 'three arquebuses, both large and small' as well as two swords, two daggers and their accessories.[6] Cellini's comments are representative of a wider fashion for and interest in firearms at the Italian courts. While Giovanni Battista Porcellaga was not giving guns as gifts, other people certainly were.

The practice of separating out weapons as a category of object in museum displays (and indeed the development of museums exclusively devoted to arms and armour) has occluded their historic incorporation into daily life and the wider court environment in the first half of the sixteenth century, a pivotal period for the firearm revolution. This chapter investigates how guns, as a new technology, were adapted to fit into the cultural world of the north Italian Renaissance courts. In the area surrounding Venice were numerous small princely states including Urbino (ruled by the della Rovere), Ferrara (the Este) and Mantua (the Gonzaga), as well as the larger polity of Milan (where control varied) and the new princely states of Florence, where the Medici (previously the leading family of the republic) ruled as dukes from 1532, and Piacenza/Parma, established by Pope Paul III as a fief for his family, the Farnese. The Italian courts formed part of a wider European social system: their dynasties allied themselves through marriage with the ruling houses of France (via Catherine de' Medici) and Poland-Lithuania (via Bona Sforza), to give just two examples. By asking how guns mingled in the apparently unmodern environment of the court, we can problematize both the modernity of the gun and the nature of the court.

Histories of the court have typically explained its trajectory as an institution from the military band of medieval monarchs to settlement in one or more key centres and a concomitant shift away from a military role (increasingly ceremonialized into the tournament form) and towards cultural activities such as art and letters. By and large, the monarchs of the early sixteenth century were the last to lead troops into battle, and often with limited success. While Castiglione conceded that the 'first profession' of the courtier remained that of arms, he added that art and letters might also grant him good reputation.[7] All this can be perceived as part of a broader 'civilizing process',[8] and historians have rightly argued for the significance of the courts in Italian historical development against an older historiography that uncritically centred republics as cradles of modernity.[9] Yet if the court society lay at the heart of a 'civilizing process' in Europe, how did the gun fit into that?

There are relatively few surviving examples of firearms prior to 1550, reflecting the limited numbers that appear in the inventories and accounts of the period; examples from the second half of the sixteenth century are far more common. Most of the armouries of the Renaissance courts were broken up following the extinction of their ruling line; that of Mantua may already have suffered fire damage in 1591.[10] The arms collection of the Este dukes of Ferrara (subsequently of Modena and Reggio) is the only one to remain in anything approaching its original form: it was transferred in 1860 to Vienna and after 1887 to Konopiště Castle outside Prague, which Franz Ferdinand Habsburg-Este used as a hunting lodge.[11] There are several large collections of arms and armour in Italy, notably at the Armeria Reale (Turin), the Luigi Marzoli museum (Brescia), the Museo Stibbert (Florence), the Odescalchi collection (Bracciano) and the Museo Correr (Venice), but they are later recreations, and in the absence of specific identifying elements such as mottos or heraldic devices on weapons it is challenging to match individual guns to original inventory references. Still, those inventories survive in sufficient numbers to permit researchers to suggest some cross-references and, more importantly, to identify the general trajectory of firearm use among the Italian nobility. Handguns appear in inventories of armouries, but also in the general wardrobe accounts of rulers, making clear that their use was not restricted to a separate military sphere.[12] Drawing on records from the courts of Ferrara, Parma, Mantua and Florence, along with examples of contemporary weapons from a range of museum collections, this chapter begins by exploring trends in elite gun ownership; it then assesses gift-giving and the processes through which firearms were effectively 'domesticated' (to borrow a term first used by John Hale), including decoration and the production of novelty weapons.[13] Overall, the chapter makes the case that the turning-point for guns' wide acceptability in court circles is to be found in the 1540s.

Ambivalence

Before turning to the specifics of gun ownership, it is worth reminding ourselves of some of the European ambivalence towards gunpowder weapons that certainly existed on the level of high culture. These attitudes went back well beyond the handgun, to the cannon of the fourteenth century. Around 1307, Armenian monk Hetoum described the Chinese thus:

> The men of theis countrey ar no stronge warryours nor valyant in armes, but they be moche subtyll and ingenyous; by mean wherof, often tymes they haue disconfyted and ouercome their ennymes by their engyns, and they haue dyuers sortes and maners of armours and engyns of warre whiche other nacions haue not.[14]

Brugh's study of late medieval and early modern German texts adds further context. Leonhard Fronsperger's 1573 *Kriegsbuch* (*War Book*) regarded gunpowder as 'indispensable', but also as 'horrible' and 'unmanly'.[15] Fronsperger had been a soldier and captain of the ducal armoury in Munich, so he was by no means an outside observer.[16] As Patrick Brugh argues, he 'worries about the loss of traditions and values, which he attributes to an imagined past inhabited by Germanic knights and manly heroes'.[17] Personal strength was no help to the warrior, wrote Fronsperger: 'It is very often the case that a manly and brave hero is brought down by a pathetic little brat with a gun.'[18] This echoes the attitude of Ariosto's epic poem *Orlando Furioso*, which described guns as a 'foul and pestilent discovery', weapons through which 'no more shall gallantry, no more shall valour prove their prowess'.[19] Fronsperger, moreover, compared the advent of firearms to a plague sent by God, a product of God's 'divine wrath' with humanity.[20] He also raised more practical issues: 'Even when we don't have the enemy in our lands any more, no one can think that he is safe from the horrible firearm on his own field or land, but he must always worry that someone will shoot at him from inside a bush.'[21]

Here his concerns parallel those of the anonymous author of the Italian treatise on the abuse of firearms, datable to the 1570s, which highlighted the threat of gun proliferation and argued for international gun control (see below, Chapter Six). These were not the only critiques of firearms. French essayist Michel de Montaigne was sceptical about their efficacy:

> a man may better assure himself of a sword he holdeth in his hand, than of a bullet shot out of a pistol, to which belong so many several parts, as powder, stone, locke, snap-hanse, barrell, stocke, scouring-peece, and many others, whereof if the least faile, or chance to breake, and be distempered, it is able to overthrow, to hazard, or miscarry your fortune.[22]

Montaigne also doubted whether rational decisions could be made in the context of warfare so dramatically speeded up by the use of firearms, especially when it came to evasive action.[23] This point finds an echo in the same anonymous Italian treatise, whose author argued that firearms had little practical utility in self-defence given the time needed to prepare them.

Moreover, at the Italian courts, where the historic environment was particularly favourable to an interest in the ancient past via either the collecting of classical artefacts or the building of classicizing buildings or both, guns were an awkward cultural fit. There was no word for gun in the Latin of Cicero so valued by Renaissance humanists. Latin poetry on the 1571 Battle of Lepanto was peppered with awkward phrasing like 'hollowed bronze' and metaphors of thunder and lightning.[24] In his account of the battle of Brescia, Ambrogio Aruscone went so far as to incorporate a mythological invention of firearms as developed by the ancient god Vulcan, though then thwarted by Juno with the rain.[25] Around 1530, another Brescian, Veronica Gambara, a poet and member of the local noble family, gave a classicizing gloss to arms production as she wrote of 'the sound of anvils ringing as the naked Cyclopes hammered out new weapons'.[26] If this worked in text, however, it did not work in the visual realm, where I have found no example of gods and guns mixing.

It is, once again, Cellini who provides a hint at the attraction of shooting outside the requirements of wartime. 'Frightened' by a plague that hit Rome in the early 1520s causing thousands of deaths, he 'began to seek relief in a sport that I found very enjoyable'.[27] In between sketching the city's ancient ruins, Cellini would shoot pigeons. He developed his own gunpowder recipe, and boasted of his 'really good marksmanship', claiming to be able to shoot 'two hundred yards point-blank'. This is much better than the test results discussed in chapter 1, and probably exaggerated, but may also indicate what might have been achieved by a highly practised shooter using a state-of-the-art gun and preparing powder and shot precisely suited to the individual weapon. Just as relevant, however, is the explanation given of what these shooting trips did for Cellini:

> Every time I went hunting my health improved remarkably from the invigorating fresh air. I have a naturally melancholy disposition, but on these trips I used to grow very light-hearted, and found myself working better and more skilfully than when I spent all my time studying and working. So all in all my gun brought me more gain than loss.[28]

In short, shooting was fun, and in a court society that prized leisure and entertainment, this was an important component in the process of domesticating weapons.

The contours of elite gun ownership

Court records illustrate a steep rise in gun ownership from the late fifteenth to the mid-sixteenth century. The 1492 inventory of Lorenzo de' Medici listed five steel arquebuses among his possessions at the time of his death.[29] By 1538/39, when an inventory was taken of Duke Cosimo de' Medici's wardrobe, the total number of firearms present was ninety-one, an eighteen-fold increase on the earlier holdings.[30] By 1561, there were 155 wheellocks alone listed, along with 52 of the larger arquebuses *da mura* (suitable for firing from a wall) and 210 *da braccia* (handheld weapons). Cosimo, whose predecessor Alessandro had been assassinated, perhaps had more concerns than most about personal protection, but the Florentine pattern is reflected in records elsewhere. Inventories document guns currently in storage, and not those currently in use, so they should not be taken as a reliable guide to total numbers, particularly not for objects like guns that might be kept by individuals for regular use in hunting. It is, moreover, important to distinguish between inventories of different departments of the court (e.g., armouries that primarily stored weapons for military use) and wardrobes that held weapons for the personal use of the ruler, his staff and courtiers (though the distinctions are not always clear-cut). Often inventories include references to worn-out guns, indicating a point in a weapon's life when it might be consigned to a wardrobe-master or armourer for repair or recycling, as we see in 1521 records from the ducal court in Ferrara.[31] An inventory of the ducal armoury taken in 1508 includes no references to guns; nor do similar records spanning the period from 1511 to 1526.[32] We know, however, that the dukes of Ferrara were deploying handguns as early as 1508 from the Modenese chronicle cited in chapter 1: most likely they were not in the 1508 inventory because they had been issued to troops. Moreover, a separate inventory of Ferrarese arms, apparently not of the armoury holdings but of weapons given to an individual named Alberto dai Morsi in 1520, includes a number of firearms, including a small *schioppetto*, a rusted gun with two barrels, a small arquebus without a stock and one old *schioppetto* in the German style, without a stock, and with a broken muzzle (*bocca*).[33] The earliest inventory among the surviving records of the Farnese court at Parma to mention firearms is dated 1540 and records two guns: a small iron *schioppo* and a second *schioppo* in the German style.[34] It also refers to four horns, presumably powder-horns, an essential accessory for the gun user.[35] An inventory of Ottavio Farnese, heir to the Duchy of Parma and Piacenza, taken in 1545, refers to a novelty wheellock (on which more below),[36] while one of the fortress at Felino (just south of Parma), dating from 1559, refers to seven 'old short arquebuses called pistols'.[37] In Mantua the pattern is similar. An inventory of 1542 lists three arquebuses, two of which are 'Bohemian' (perhaps gifts from the eastern lands of the Holy Roman Empire), along with accessories, eleven bronze *schioppi* and a novelty combination crossbow 'with four *schioppi* inside'.[38] A parallel inventory of 1543, incorporating additions through to the 1560s, illustrates the changes to these holdings, notably the receipt and giving of arquebuses as gifts in 1547, 1553, 1562 and 1567 (discussed below), and the receipt of gun parts (barrels and wheels) as well as, in 1566, the deposit of twenty-eight handguns of which twenty-one were wheellocks, an indication of the permeation of this contentious technology.[39] The slow increase in gun ownership at court between the 1520s and 1540s suggested by these documents corresponds with the period of transition towards firearm use in wider Italian society, and in particular the establishment of militias.

Wheellocks

The presence of wheellock firearms in the court records is notable. Expensive, convenient and modern, they seem to have become a status symbol. As we have seen, the standard military firearm of this period, the matchlock, was fired by means of a lighted cord (the match) which ignited the powder when the gun was triggered. The wheellock was a self-lighting mechanism that avoided the need for a lighted match. It was a more complex and therefore more expensive technology and could be unreliable (probably especially if produced on the cheap). On the other hand, wheellocks had advantages in hunting, avoiding as they did the glow and smell of the burning match cord which might alert prey to a hunter's presence. However, the fact that wheellocks could easily be concealed prompted concerns around social order. Both German and Italian origins have been hypothesized for the wheellock; given the regular exchange of personnel between the two regions, this does not need to be an either/or. One study has pointed to the figure of Giulio Tedesco, a German craftsman in the household of Leonardo da Vinci, as a possible point of connection.[40] Leonardo's sketches of wheellock mechanisms can be seen in the Codex Atlanticus.[41] German expertise was certainly important in training early firearm users in Italy, including the militia of Florence, established in 1506 (see chapter 5). In any case, wheellocks remained legally in use throughout the period among elite bodyguards and in cavalry contexts (being

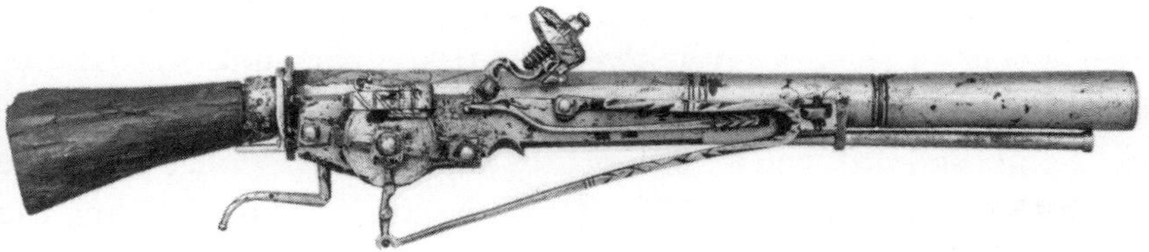

FIGURE 4.1 Unknown maker, wheellock pistol, c. 1520. Pontebba, Italy. Steel, wood. 39.4 cm length. Royal Armouries, XII.1765. © Royal Armouries.

more practical for use on horseback than the matchlock, which required the use of both hands). A wheellock may have been used for hunting by a member of Cardinal Ascanio Sforza's entourage in or before 1503; another early record comes from 1507, when Cardinal Ippolito d'Este (brother of military technology enthusiast Duke Alfonso of Ferrara) purchased a weapon that could be 'fired with a stone'; they may have been used by cavalry in the 1510s.[42] A further example, notable for its open mechanism and perhaps dating as early as 1520, is to be found in the collections of the Royal Armouries (Figure 4.1). Found near the Hungarian/Slovak border, where it had been preserved in a bog, it is similar to examples from the Venetian *terraferma* (mainland).[43] Italian troops campaigned in this area in the 1520s and 1530s during the extended contest for hegemony in Hungary between the Holy Roman and Ottoman Empires; there was also considerable exchange of personnel and gifts between the courts. Other early examples, including combination weapons, are to be found in the Palazzo Ducale, Venice, and the Landeszeughaus, Graz.[44]

By the start of 1538/39, Duke Cosimo de' Medici owned four wheellock handguns made in the German style which he took with him when he travelled.[45] Presumably these were carried by his bodyguard. More generalized use of wheellocks seems to have taken off in the 1540s. The duke of Urbino, who regarded the gift of a firearm as 'very dear', was in 1551 permitted by the Venetian authorities to arm his bodyguard with small wheellocks, although these were only to be worn when accompanying him.[46] Updates to the legislation suggest that many courtiers thought they should likewise be entitled to make use of the new technology, but that rulers were not always keen to extend such privileges. A 1551 ban on wheellocks in Ferrara explicitly stated that it also applied to members 'of the court of His Excellency and the Most Illustrious Madama and the Lord Prince', unless they had 'express licence' from the duke.[47] This was a wartime measure, introduced shortly before the papal siege of the Ferrarese city of Mirandola, and there may have been particular concerns about assassination or internal rebellion. In the longer term, wheellocks continued to proliferate.

Guns as gifts

The importance of gift-giving at the Renaissance courts is well established. Guns became a part of that gift culture, given to strengthen bonds between ruler and courtier, patron and client, or in the context of diplomacy. The most prized guns were by no means exclusively of Italian production. The inventories are notable for the number of German and Imperial-style weapons listed. (Imperial in this context refers to the Holy Roman Empire.) This reflects in part the earlier establishment of a firearms industry in southern Germany, which began to specialize in luxury weapons prior to the growth of such production in Italy (it is only from the seventeenth century that substantial numbers of decorative Brescian weapons survive). A 1558 report from a Venetian ambassador to an Imperial court noted that there wheellocks were 'carried by everybody', a more open gun culture than the prevailing one in Venice, which tightly regulated such weapons.[48] The presence of German firearms also reflects the growing power of the Habsburgs in Italy in this period when Charles V ruled both Spain and, as Holy Roman Emperor, much of what is now Germany and Austria. Alessandro de' Medici, duke of Florence from 1532 to 1537, gave a new German *schioppo* to the lord of Monterotondo, father of one of his courtiers, and a German matchlock gun to his chief minister Cardinal Innocenzo Cibo. The timing of the latter gift suggests the gun may have come from someone in the entourage of his fiancée, Margaret of Austria, who had recently visited Florence. Margaret (more commonly known as Margaret of Parma following her second marriage to Ottavio Farnese) was the illegitimate daughter of Charles V. The gun may even have come from Margaret herself: her aunt, Queen

Anna of Bohemia, hunted with a firearm in the 1520s.[49] The earliest surviving wheellock weapon that can be linked to a specific individual is a combination gun-crossbow belonging to Anna's husband Ferdinand (dated between 1521 and 1526); wheellocks dating to the 1530s can be found in the Spanish royal armoury.[50] The presence of German commanders and diplomats in the service of Charles V no doubt facilitated such transactions, as did connections via the numerous Habsburg women who from the 1530s onwards became the brides of Italian rulers (Margaret, duchess of Florence from 1536 and later of Parma; Catherine, briefly duchess of Mantua in 1549–50; Eleanor, duchess of Mantua from 1561; Barbara, duchess of Ferrara from 1565; Johanna/Giovanna, grand-duchess of Tuscany from 1565).[51] In 1549, Giovanni Battista da Gambara, member of a prominent Brescian noble family (one not always in accord with the Venetian overlords) wrote from Venice to the duke of Mantua to recommend the work of a German wheellock maker who 'as a boy' had been in the service of the duke's father. This man was now a 'good master', whose service Giambattista would happily secure for the duke: he was 'young, bright and works willingly'.[52]

These personal relations are echoed in the material record. There was an old *schioppetto* in the German style (*ala todesca*) at Ferrara in 1520, and a German *schioppo* at Parma in 1540.[53] Such were the numbers of German wheels at the Florentine court by the 1560s that they had a separate listing in the wardrobe books.[54] In the context of growing Imperial power in Italy, this terminology may have carried some symbolic weight as well as having a descriptive function. I have come across no corresponding French examples in the north Italian inventories, although they did exist: the Metropolitan Museum of Art holds a matching pair of sixteenth-century French wheellocks plus flask and spanner.[55] The Imperial producers seem to have dominated the international luxury market at this point. There were similar patterns in Mantua. Two arquebuses 'in the Bohemian style' were present at the court there by 1542; in 1547, the duke gave one of the Bohemian arquebuses that had been in his wardrobe four years before to Contino da Gambara, along with a gilded morello velvet flask and further accessories.[56] In 1548, the duke gave a small *schioppo* barrel to the son of his stablemaster, a gift notable for the implication that the recipient would have been able to have it made up into a full weapon.[57] From 1553, the podestà of Mantua (officials responsible for security, an appointment traditionally held by a non-citizen) began to make ceremonial presents of firearms to the duke in place of the crossbows they had previously given.[58] This shift indicates that by mid-century any negative associations of firearms had largely dissipated in favour of a court culture in which they were generally welcomed. The process had taken place over several decades, but by this point it must be regarded as close to complete.

By now guns had also entered the realm of the diplomatic or political gift. In 1539, Cosimo de' Medici's secretary wrote to Count Pier Francesco da Noceto, noting that the count's messenger would be able to advise him how much his gift of an arquebus had pleased Cosimo.[59] In 1551, Don Diego Hurtado de Mendoza, Spanish governor of Siena, sent a gift of an arquebus to Cosimo, one of many that the duke received.[60] On 12 April 1556, Cosimo was gifted three arquebuses, one *schioppetto* and a flask by the authorities of the Hospital of Santa Maria Nuova; later the same year he received six Turkish arquebuses, perhaps spoils of war, from Marco Centurione, a Genoese naval commander.[61] He appears to have gifted some of these in turn because the records note that one long Turkish arquebus was to be sent to the cardinal of Mantua (Ercole Gonzaga, who as a younger man was known for his love of hunting); two further long Turkish arquebuses were to be dispatched to Count Rados and to Balena, an arquebusier.[62] Similar guns are listed in an Orsini inventory of the late 1570s or 1580s.[63] War is not the only possible route for transfer of these weapons, however: by this time a Franco-Ottoman alliance was well established and the Italian states were long-standing importers of all manner of goods from the Ottoman Empire. A surviving German wheellock of 1540 shows an Ottoman archer on horseback on its stock, and Ottoman sources of the period provide testimony to the use of firearms in the military campaigns of Sultan Süleyman the Magnificent.[64]

Domesticating guns

Alongside their assimilation into the gift culture of the court, guns were also assimilated into its decorative environment and thus 'domesticated'.[65] We have already seen how imagery of firearms on *cassoni* panels placed the use of these weapons into a domestic environment. However, the decoration of guns themselves also mattered. Indeed, the weapons that were given as gifts were often as visually attractive as they were useful (and in some cases probably more so). Decorative continuities with broader court culture helped guns blend into the wider court environment, while novelty weapons appealed to an appetite for fun and games. While the majority of guns had plain wooden stocks, luxury variations incorporated gilding, intarsia work and materials including ebony, ivory and (for accessories) velvet: all familiar from other domestic objects. A variety of woods—some more prestigious than others—were available for production of the stock. Designs echoed imagery common to the court, with themes of hunting or battle that might equally be found on tapestries. Inventories rarely provide precise descriptions of the designs, but these may be inferred from the variety of surviving weapons, some of which can be linked to specific owners and producers. In the introduction to the catalogue of firearms of the Stibbert Museum, Florence, Riccardo Franci writes that 'it was at the beginning of the sixteenth century that firearms began their dual life as useful (for military, defensive, or hunting purposes) instruments and works of art'.[66] This distinction, however, needs to be clarified: it was necessary for guns to become works of art in order to normalize their use in other contexts. As objets d'art, guns suffer from comparison with such items as parade armour, the large surface of which offers wider scope for elaborate design and thus more scope for appreciation among the decorative arts. As Springer has shown, as armour lost its battlefield utility over the course of the sixteenth century it acquired a distinct symbolic role in the court environment.[67] Even at smaller scale, however, guns could be inlaid and gilded, and their designs could communicate a range of cultural messages.

Decoration was one means by which firearms could be incorporated into an environment that prized the classics. If guns were an awkward weapon in terms of options for description in Latin, they could readily be decorated with motifs from ancient history and myth. A South German powder-flask now in the Metropolitan Museum of Art includes a scene of the Roman general Scipio storming New Carthage (Figure 4.2).[68] An ivory wheellock gun

FIGURE 4.2 Unknown maker, powder-flask with bullet box, clock, compass and sundial, c. 1570–1600, Augsburg or Nuremburg. Brass, gold, steel, glass. 20.3×11.4 cm. Gift of William H. Riggs, 1913. Metropolitan Museum of Art, 14.25.1493. Public domain.

belonging to Philippe de Croy, prince of Chimay (also in the Met) features a design of the ancient hero Perseus.[69] Guns sometimes showed women associated with conflict like Lucretia (whose rape and subsequent suicide became part of the founding myth of the Roman Republic) or alluded to the Trojan War (allegedly begun by Helen, whose face launched the thousand ships).[70] The biblical figure Judith (who saved her people from invasion by beheading the enemy general Holofernes) also commonly features, although some of the female figures on guns are less symbolic, such as the Venus leaning on a lyre featured on a gun from around 1570 owned by Charles II, Archduke of Inner Austria.[71] Scenes of courtship and embracing couples are also frequently found: these depictions of women are arguably an early equivalent of the 'nose girls' who in the twentieth century were painted on military aircraft.

By mid-century, in a break from descriptions of guns as diabolical, it had become more common to combine classical and Christian elements. One notable firearm from the 1540s is the wheellock pistol now in the Odescalchi collection, of German manufacture (probably Brunswick) and dated 1548. Its barrel features a figure of Lucretia but her classical image is combined with an invocation to God: 'BESHUTZE MICH HERR GOTT UND RETTE MICH VAN MEINEN FEINDEN' (Lord God protect me and save me from my enemies), and the butt of the stock is adorned with an image of the Crucifixion. The wheellock plate is decorated with foliage and medallions while the stock is inlaid with bone and again incorporates classicizing and mythological elements.[72] A similar combination is to be found on the 1555 pistol belonging to Ferdinand II, now in the Kunsthistorisches Museum, Vienna (Figure 4.3). It incorporates foliage designs and hunting imagery along with a depiction of the Crucifixion and one of Mucius Scaevola (a Roman youth famed for his attempt to assassinate an Etruscan general and bravery when captured). It is no surprise that this imagery appears at a point when guns were finding new uses in Europe's religious wars: far from being the Devil's weapon, they were now a means of asserting true Christianity against the heretics.

Hunting scenes were also common in gun design, as were armorial devices or other allusions to dynastic power. Among the weapons of Cosimo de' Medici was one with an ivory and black stock featuring the ducal arms.[73] The surviving part of the Este collection now at Konopiště Castle outside Prague includes a very similar gun dated 1544 and decorated with the Medici arms and inscribed 'COSMS MEDIES REIPIES FLORE DV' (Cosimo de' Medici, duke of the Florentine Republic). The barrel of this wheellock is inlaid with gold and silver in floral designs, while ivory plates covering

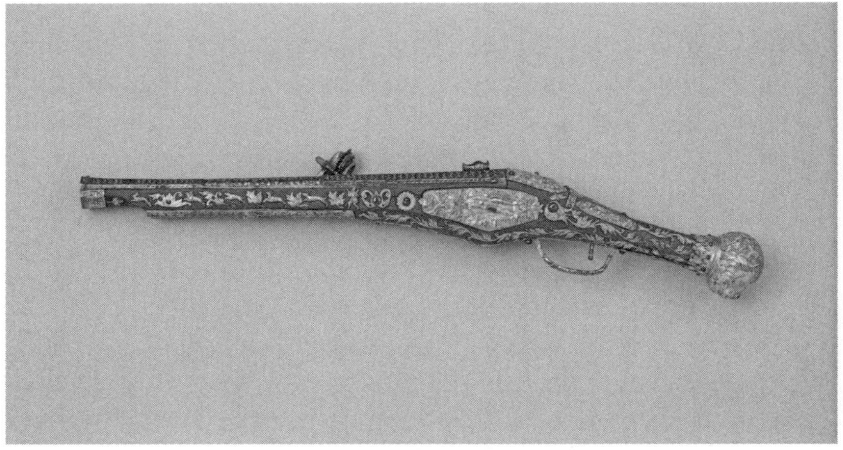

FIGURE 4.3 Unknown maker, wheellock pistol for Archduke Ferdinand II of Tyrol, c. 1555, Brunswick. 64.5 cm length. Iron, silver and gold inlay, gilt, walnut, other wood. KHM, Hofjagd- und Rüstkammer, A 525. © KHM-Museumsverband.

the stock incorporate scenes of hunting and the image of a gun crossed with the fasces representing ducal rule.[74] This striking (and provocative) statement about sources of power is, however, concealed on the underside of the weapon. A further gun (*schioppo*) listed in Cosimo's inventory was worked with gold and white and had an ivory stock featuring 'battles and hunts'.[75] Dynastic images are also to be found on Charles V's double-barrelled pistol, made in Munich by the gunsmith Peter Peck and etched by Ambrosius Gemlich, which incorporates Charles's insignia, the double-headed eagle and the Pillars of Hercules, as well as the motto 'Plus Ultra', an allusion to the mythical inscription 'Ne Plus Ultra' on the Pillars at the Strait of Gibraltar (Figure 4.4).[76] This gun, too, is decorated with a hunting scene, perhaps featuring Charles himself. Another German weapon of the 1540s features the Judgement of Solomon, along with a variety of emblems including the *biscione* (the heraldic device of a serpent swallowing a man), which might suggest a connection to Milan.[77] The inventory of Paolo Giordano Orsini likewise includes mention of a gun featuring the family arms.[78]

If design was one means of assimilating a firearm into an established court environment, so were materials. The wheellock pistol dated 1545–55 and currently in the Correr collection provides

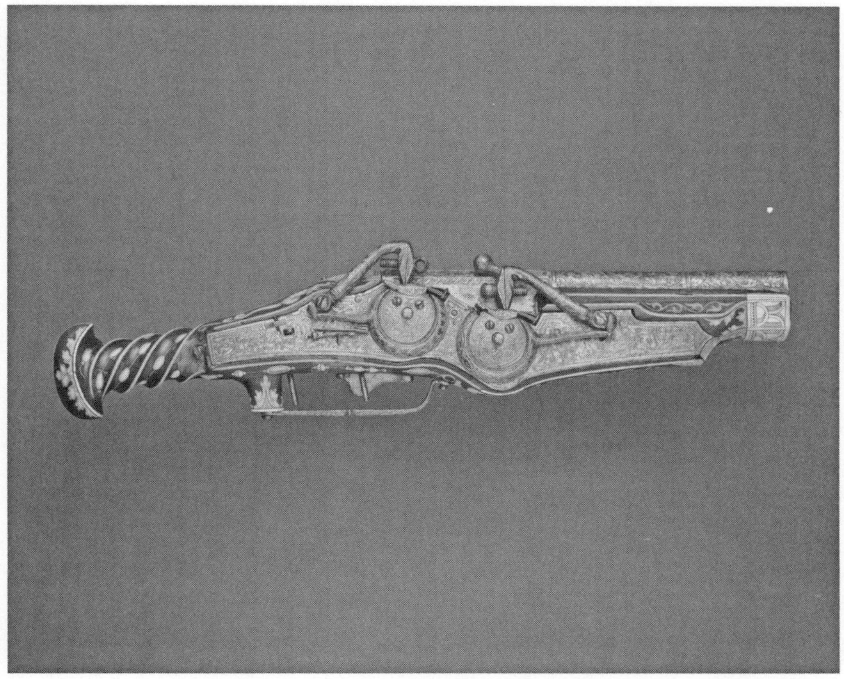

FIGURE 4.4 Peter Peck (gunsmith), Ambrosius Gemlich (etcher), double-barrelled wheellock pistol made for Emperor Charles V, Munich, c. 1540–45. Steel, gold, wood (cherry), staghorn. 49.2 cm length. Gift of William H. Riggs, 1913. Metropolitan Museum of Art, 14.25.1425. Public domain.

a good starting point for the design of such weapons.[79] Just under thirty-six centimetres long, its barrel features acquaforte decorations of foliage (engraved onto the metal with acid), while its stock is inlaid with ivory in a design of scrollwork and foliage. The wheel is set on a gilded plate featuring an acquaforte design incorporating dense foliage and an eagle. A small breech-loading pistol decorated with bone and now in Vienna may be dated to c. 1540 and offers an early, subtle example of such inlay, while ivory and bone flasks survive too (Figure 4.5).[80] Gilding was frequently used to decorate luxury weapons. A Florentine inventory of 1553 records two new German wheels 'touched with gold',[81] as well as two small gilded arquebuses and six further German wheels, 'two gilded and four worked with azure, all most beautiful'.[82] Both gilding and azure may be found on daggers listed in the wardrobe: the decor here

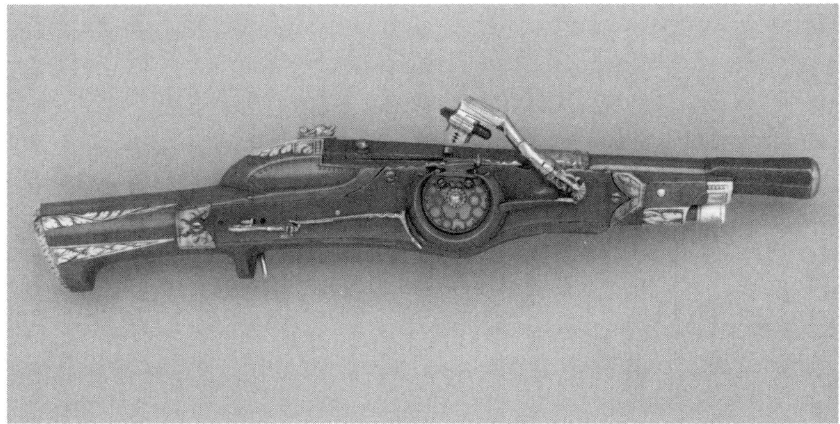

FIGURE 4.5 Unknown maker, wheellock breech-loading pistol, c. 1545, ?Augsburg. Iron, bronze, wood, bone, 46 cm length. KHM, Hofjagd- und Rüstkammer, A 603. © KHM-Museumsverband.

reflects continuities with other weapons. These decorative techniques also, however, provide more general continuity with furnishings and clothing. Intarsia work featured on chests and tables; ivory was incorporated into the most prestigious examples.[83] Gilding was an expensive way to finish furniture, but consequently a means to demonstrate wealth.[84] Cordovan leather, used to produce accessories for firearms, was likewise particularly prized amid a broader fashion for Spanish gilt leather wall-hangings.[85] In short, an inlaid weapon might resemble the surface of a card table; its decor echo the material on the wall behind its owner.

To explore these material continuities, we might consider the house of the Bardi and Cavalcanti firm in London. This company, which had close links to the Medici family, was engaged in the import of multiple products, including guns, to the Henrician court. The 1523 inventory of its London house included eleven tapestries, of which nine featured designs of *verzura* (vegetation) and/or animals, while two featured figurative historical themes. The artwork was predominantly religious, but one canvas of a Roman hero, Marcus Curtius, also featured.[86] Marcus Curtius in turn appears on the 1570 gun owned by Charles II of Inner Austria, made by Hans Paumgartner.[87] The decorative themes likewise blend into the court environment.

Foliage designs were popular on tapestries, as were hunting scenes. Religious and classicizing imagery were routinely to be found in court environments. While the particular images of Lucretia or the Judgement of Paris (seen on a 1571 gun also by Paumgartner and probably made for the wedding of Archduke Charles of Styria) have obvious links to warfare, they are hardly unknown as decorative elements.[88] The insertion of coats of arms and other heraldic devices likewise follows general practice in courtly culture, where family insignia were incorporated on all manner of domestic objects.[89] The fact that the parts of a gun could be produced and shipped separately opened up opportunities for customization, allowing a stock made to a patron's design to be attached to a barrel and lock made by a specialist. The Medici wardrobe accounts show that on 15 February 1553/54 'Francesco legnaiouolo' (Francesco the carpenter) was given on the orders of the duke two new arquebus barrels for shooting birds so that he could make cherrywood stocks for them.[90]

These decorative approaches were carried over to the accessories for luxury guns, although the design continuities here are less with the domestic interior and more with fashionable clothing. The Medici inventory of 1560, for example, records 'five flasks of black velvet and a *polverino* [priming-flask] and holster'.[91] Two large arquebuses, of the type fired from a rest, were accompanied by sheaths of red and black leather: leather was widely used to make these early holsters. One purse (perhaps for shot) was decorated with black velvet cord and tassels.[92] A combination of black velvet and gilding would have fitted very well with the fashion for Imperial black set by Charles V who was known for his expensively dyed monochrome wardrobe.[93]

Experimental and novelty weapons

There were other means, too, by which firearms might be aligned with the values and interests of the court. As we have seen, samples were often provided in advance of a large military order: novelty weapons were an extension to that practice. Moreover, they tied into a fashion for wider theatrical and festival machines aimed at provoking wonder in their audience. These were widely displayed at Renaissance courts and include such things as the mechanical lion designed by Leonardo da Vinci for the entry of the king of France to Lyons in 1515.[94] In the early seventeenth century, a fountain designer proposed to adapt one of the water features in the gardens of the Villa d'Este in Tivoli, near Rome, to incorporate two statues of arquebusiers that would fire water at visitors.[95] Guns involve an element of spectacle: the wondrous ability to produce fire without a lighted match, for example. They fit the sixteenth-century Italian fashion for entertaining domestic objects, such as novelty mugs for drinking games.[96] The Este collection now at Konopiště, for example, includes an arquebus with both wheellock and matchlock mechanisms, attributed to Wolf Danner of Nuremberg.[97] Danner, who may also have produced a wheellock rifle for King Gustav Vasa of Sweden, was one of the most prominent court gunsmiths of the period, alongside Peter Peck, who made many guns for Charles V.[98] Triple-barrelled firearms survive in several Italian collections, among them a distinctive German wheellock example with rotating barrels (incorporating the arms of Charles V) now in Turin.[99] The presence of such weapons in mid-century Italy is confirmed by the reference to a small double-barrelled wheellock arquebus in the 1553 wardrobe accounts of Duke Cosimo de' Medici.[100] A further inventory taken in 1560 records several additional firearms of this type: two small triple-barrelled wheellock arquebuses; one arquebus with three wheels that produced three shots; and a further double-barrelled arquebus.[101]

More extravagantly eccentric were the many combination weapons that have often been retained in collections. Such firearms appear frequently in inventories, which may reflect the lack of ongoing utility and consequent storage following an initial demonstration. The Royal Armouries in England holds a match-fired combination poll axe and gun, dating from the early sixteenth century.[102] Henry VIII, who as we have seen imported guns via an Italian company, owned a number of such weapons at his death in 1547.[103] The two combination wheellock axes now in Venice (but probably German) dating to around 1515 may be the earliest surviving weapons with a wheellock function.[104] A later combination axe-pistol, first recorded in the 1589 inventory of Grand Duke Ferdinando de' Medici and featuring the family's arms, is in the Metropolitan Museum of Art (Figure 4.6). The Met collections also

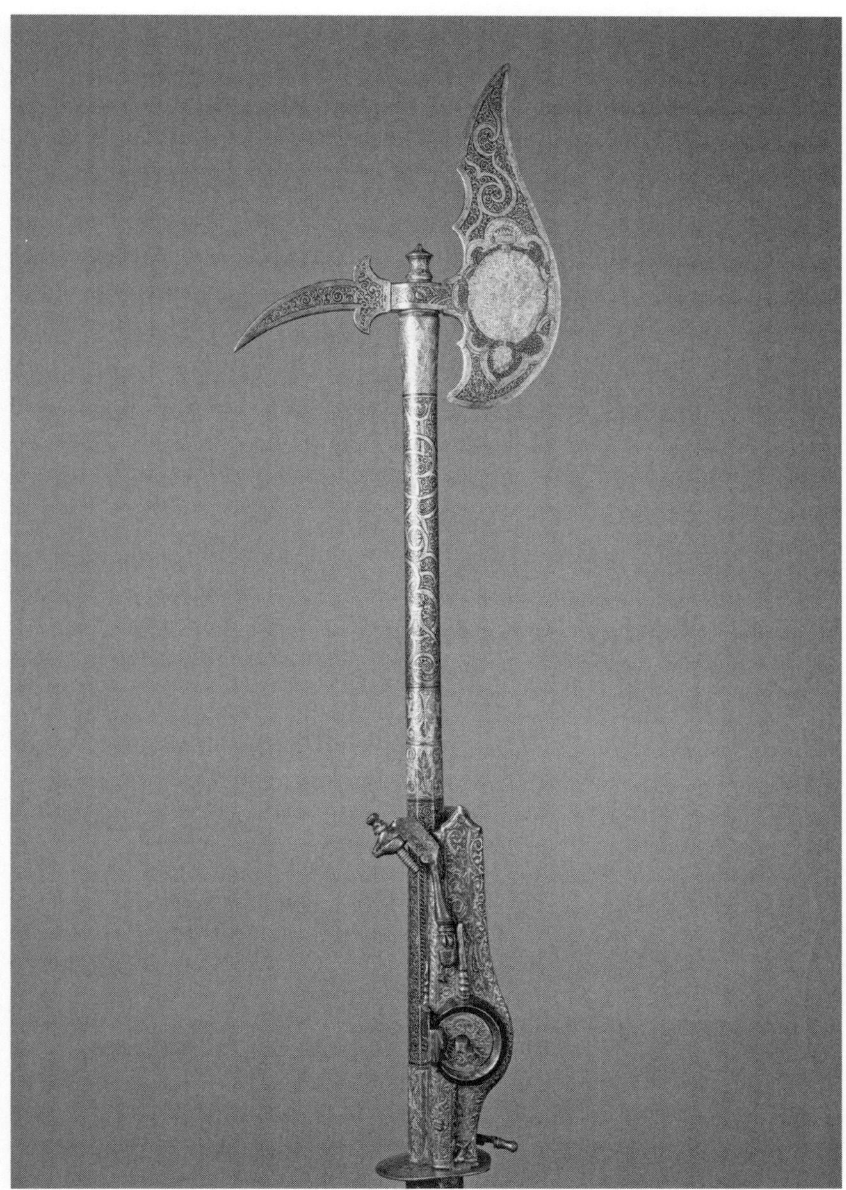

FIGURE 4.6 Unknown maker, combination axe-pistol of Grand Duke Ferdinand I de' Medici. Germany, c. 1580. Steel, gold. 70 cm length. Purchase, Arthur Ochs Sulzberger Gift, 2002. Metropolitan Museum of Art, 2002.174a, b. Public domain.

include a hunting knife with a wheellock pistol (1528–46) and two maces with wheellock pistols (c. 1550).[105] The 1543 Mantuan inventory lists an iron axe with two *schioppi* and an iron crossbow with four *schioppi* inside.[106] In 1566, Cosimo de' Medici received from the incoming commander of his German ducal guard a boar-spear with two small arquebuses attached.[107] This would not have been a great novelty because the 1561 records show that he already owned such an item, inlaid with gold and with a leather sheath, along with three more boar spears each containing a single arquebus.[108] Perhaps the commander was playing safe with a gift he knew was to his new employer's taste. A similar item is recorded in an inventory of Paolo Giordano Orsini, duke of Bracciano 1560–85; a surviving example, dating from 1560 and owned by Archduke Ferdinand II of Tyrol, is now in Vienna.[109]

As the technology developed, guns were fitted into smaller and smaller objects. The 1545 inventory of Ottavio Farnese, for example, lists a small gilded wheellock arquebus 'in the style of a dagger handle'.[110] A similar weapon appears in the Medici wardrobe accounts of 1553: this is a 'little gilded arquebus in the guise of a dagger'.[111] These accounts also list two similar small arquebuses 'in the guise of a set of knives' which came in sheaths of velvet and black leather.[112] A subsequent inventory, taken in 1560, explains that each set contained four knives alongside the disguised gun, and that one set had ebony handles while the other had bone handles; they were described as *alla todesca*, in the German style.[113] Records from the following year give more detail of a pair of small wheellock arquebuses 'in the style of a dagger', which were worked *alla damaschina* (i.e., with metal inlay) and came with two small flasks, one of white bone decorated with foliage and one similar in black, plus a powder-flask of painted iron.[114] Combination accessories also existed: the collections of the Luigi Marzoli Museum in Brescia include a powder-flask incorporating a clock.[115] These weapons functioned to some extent as marketing devices, promoting their makers—and perhaps also the cities or states of their makers—as potential providers too of the thousands of weapons that might be commissioned for use in war. While as weapons they may seem eccentric, they fit very well into the context of broader court culture, where trickery and concealment were common not only in cultural environments such as the masquerade but also in political practice. A gun disguised as a dagger handle is on one level a simple extension of the fashion that might put blackbirds in a pie to surprise unwary banqueters. However, as with decorative styles, the practice of placing firearms into this environment served to normalize them. Guns were thus assimilated into everyday noble culture, linked to themes such as novelty, deception and the hunt.

Conclusion

Handguns were undoubtedly present at the Italian Renaissance courts: if some ambivalence about them remained, then it did not exclude their being given as gifts. Indeed, it was vital to the firearm revolution that guns should gain such acceptance. While in cultural contexts firearms were often represented as antithetical to a chivalric ethos, in fact those people with the best access to the most effective weapons for killing, and an appreciation of their quality, were precisely the noblemen and patricians who regularly pleaded for exemptions from the usual bans and penalties on behalf of their citizens.[116] Novelty weapons might entertain guests in the manner of any practical joke; those made from gold and silver literally embodied their owner's wealth. The decoration and materials of luxury firearms aligned them with the existing court environment, though positive writing about them remained rare. Cellini, in any case as an artist on the margins of court life and writing an edgy swashbuckling account of himself, was an exception. Francesco Maria della Rovere, whose enthusiasm for wheellocks we saw at the beginning of the chapter, was captain-general (commander-in-chief) of first the papal armies and then the Venetian; as such he might be expected to take an interest in developing technologies. He is also, however, known for his personal violence which even in his own time prompted comment: in 1511, he killed Cardinal Alidosi, and he was later involved in the commissioning of other murders.[117] Those facts offer a rather different gloss on his interest in lethal technology, reminding us that even in the court environment violence was never far away.

By considering firearms within the world of courtly material culture, rather than artificially consigning them to a separate military sphere, we gain insights not only concerning the role of technology but also into the nature of the court itself. There is no reason to be surprised at the connection: it has long been noted that the military contracting system helped redistribute wealth from larger states to smaller ones, providing their princes with the resources for art commissioning.[118] This exploration of guns in court context adds cultural nuance to narratives of state formation and in particular the rise of the fiscal-military state, demonstrating how violent technologies were incorporated into the material world of the rulers who subsequently sought to monopolize that violence.

Moreover, attitudes towards firearms should be contextualized in relation to broader changes in court life at this time. Monarchs were less and less involved directly in warfare. The court was becoming a centre for government, and increasingly a settled one. Events such as the tournament were less and less a matter of practical training and more a matter of symbolic performance. In this heavily ritualized and symbolic environment, the practical use of guns on the battlefield could be set aside, and shooting become sport. If a part of the social elite now had privileged access to pocket-sized lethal weapons, rules of civility were meant to ensure that they would not be used. Over time, an idea of the gun as itself a more civilized weapon—one that avoided physical contact and maintained distance—would develop.[119] In the sixteenth century, that thinking was still some way off, but the assimilation of guns into the court environment with its growing emphasis on etiquette, bodily control and propriety was one part of the process. More generally, as we will see, wider ownership of firearms went alongside a process of social disciplining that—albeit with difficulty—restrained the use of this new technology.

CHAPTER FIVE

Gun Proliferation

WIDER SOCIETY WAS as ambivalent about handguns as the courtly elite, seeing them as both useful tools and social problems. Over the course of the sixteenth century, however, as the firearm revolution played out, guns became a routine object in European lives. In this chapter and the next, I explore first how they proliferated, and then how states—sometimes rather late in the day—sought to regulate them. The contrast between orderly and disorderly uses of guns is apparent in a fresco by Giovanni Stradano (Jan van der Straet, a Medici court artist) on a staircase in the Palazzo Vecchio, Florence. Dating from around 1558, it shows a firework display held in the square outside, Piazza della Signoria, to celebrate St John's Day (Figure 5.1).[1] Explosives were a regular feature of Florentine civic celebrations and, besides the fires burning in the square, the fresco shows a display of shooting from arquebusiers lined up along the wall of the palazzo. However, a handful of small figures in the foreground are also firing into the air in an apparently less organized fashion. In normal circumstances, guns were not permitted within the city of Florence. Perhaps these were members of the ducal guard with licence to shoot, or perhaps the piece is an imagining on Stradano's part. Either way, the fresco shows the two sides of guns in an everyday city environment.

This chapter looks beyond the world of luxury weapons and the court to consider the routes through which firearms proliferated. Besides export licensing, allowing some civilian access to firearms was one way for states to ensure there was a market for arms

FIGURE 5.1 Jan van der Straet / Giovanni Stradano, *St John's Day Fireworks in Piazza della Signoria* (detail), c. 1558. Fresco, Palazzo Vecchio, Florence. Photo: Azoor Photo via Alamy.

producers in times of peace, and guns became familiar objects at all levels of society. They were legally used by, and indeed required for, members of the militia, a long-standing European institution but one that gained a new impetus in the context of the Italian Wars. They were employed in hunting and for livestock control; they were carried for personal security while travelling. Firearms were also put to illegal uses, especially by the bandits who became notorious in later sixteenth-century Italy, but also by those engaged in political violence, categories that cannot be entirely separated. The chapter explores these issues in turn, concluding with a discussion of what we know about ownership of guns. This sets the scene for the following chapter, which considers gun control in greater detail.

Civic militias

The single most important factor in the proliferation of firearms in sixteenth-century Italy was the development of civic militias.[2] Militias were not new: the 'golden age' of the Florentine militia, according to C. C. Bayley, lay in the thirteenth century,[3] though debates on the subject continued in the city through the fourteenth and fifteenth, with contributions from future city chancellor Leonardo Bruni in 1422 and polemicist Domenico Cecchi in 1496.[4] By the 1490s, large-scale foreign military intervention on the Italian peninsula was giving a new impetus to militia development, as was the growing importance of infantry in the military context. In Friuli, where Venetian lands bordered Ottoman territory, the militia were using handguns from the 1490s, and city guards across the Italian peninsula were soon provided with guns too.[5] The provision of 1506 that established the Florentine militia required that one in ten men should be equipped with firearms and that five hundred handguns (*schioppetti*) should be retained in the Palazzo della Signoria (the city hall, now known as the Palazzo Vecchio) along with 'such money as should be necessary for the *schioppetti*', presumably in order to purchase gunpowder and shot. Three or four companies (*bandiere*) entirely of *schioppettieri* might also be established.[6] This militia, most often noted for the role of Niccolò Machiavelli in its foundation, was to be organized after the fashion of the Germans, in whose cities there was a centuries-old tradition of civic defence and a duty on every male citizen to maintain suitable arms for the purpose.[7] By the end of the sixteenth century, many of the suitable arms would, in fact, be guns.

In the context of the Italian Wars, a melting-pot of military expertise, those establishing militia also took inspiration from the Low Countries and Switzerland, where the use of guns was integrated into a wider martial culture. For many medieval towns, military service had been a quintessential element of urban identity, a feature sometimes obscured by the emphasis on the role of the state in making war.[8] These civic militia systems led to widespread arms ownership in cities. In the Low Countries, where the tradition was particularly well established, shooting guilds (using crossbows) dated back to the thirteenth century. The night watch ordinance of Basel from 1411 required watchmen to be armed: at that point the usual weapons were a battle axe or halberd.[9] The wealthy city of Ghent was among the first to establish an artillery park, and had a firearms guild by the end of the fifteenth century.[10] Vienna had a long history of civic obligations to participate in military activities: citizenship fees could be paid with a contribution of weapons or armour and between 1468 and 1504 there are records of some individuals, albeit fewer than 5 per cent of the total, making their payment in arquebuses.[11] Urban militias were not necessarily counterposed to ducal or princely armies but indeed served within them: cities including Fribourg (Switzerland) had such militia companies.[12]

Developments in Florence may also have been inspired by the reorganization of the papal defences on Spanish lines undertaken under Pope Alexander VI, whose son Cesare Borgia Machiavelli had met in the course of diplomatic negotiations and evidently admired.[13] Indeed, while one of the Florentine militia's early constables was German (a Giovanni Tedesco), another was Spanish (Giliberto Spagnolo). Giovanni Tedesco arranged target practice for his *schioppettieri*, which from 1508 was extended more generally among the militia, with exercises taking place on a monthly basis.[14] Arrangements for militia drills were made in Cagliari, Sardinia (then under the rule of a Spanish viceroy), in 1528.[15] The 1520s also saw the extension of militia training in Venetian territory, although reports of the Siege of Padua in 1509 suggest that there was already considerable informal knowledge of shooting among the peasantry beyond the Friuli borderlands.[16] Giovanni Moro, a former *provveditore in campo* (the official responsible for supplies to the army camp) and rector of the subject city of Crema, proposed in 1525 the formation of a citizen militia of twelve thousand men, of whom a third would be arquebusiers and a further third *schioppettieri*; these men would receive powder and lead for practice, and a contest would be held three times a year to encourage them.[17] This would be, said Moro, to the 'honour and utility' of Venice, and of 'great benefit' when it came to conserving Venetian territory.[18] Venice's dramatic loss of most of its mainland holdings in 1509, and subsequent efforts to recoup them, provided the longer-term context for these debates.[19] In 1528, the total size of the Venetian arquebus militia across all territories was noted by a city senator, Marin Sanudo, to be twenty thousand men.[20] However, we know that by March 1530 the total number sought from Brescian territory alone was four thousand, so twenty thousand in total may be an underestimate.[21] In any case, these numbers show the considerable efforts made by governments to expand the militia and with it knowledge of shooting.

Such ambitious plans did not always succeed, as two cases from Tuscany show. Jacopo Pessina's detailed study of military organization in the Republic of Siena estimates that in the last thirty years of the republic (1525–55), between 20 and 33 per cent of adult men served in the militia, with each 500-strong battalion including 150–200 arquebusiers. A report from 1543, however, suggested that the militia was not fit for purpose, lacking not only sufficient men but also the necessary firearms and helmets.[22] The case of Florence, moreover, where the last republican government of 1527–30 fell to the Spanish-backed Medici party after a long siege, shows the limits of a city militia in the face of a foreign army. Anticipating attack, the republican government had issued new ordinances for the militia, instituting training in the use of arquebuses and shooting contests.[23] Lists drawn up by the Otto di Pratica (Eight of Practice, the magistracy responsible for external and military affairs) during the siege in 1530 show that artisans from shoemakers to cutlers were equipped with arquebuses.[24] They did not, however, prevail against the much better-resourced Spanish army.

Still, governments continued to expand their militia, which made firearms increasingly available to men of all social ranks. By 1550, there were 6,463 militiamen with guns in the Florentine dominion.[25] In October 1552, anticipating an attack on the city, the Sienese authorities ordered the purchase of a thousand arquebuses from a merchant in Pistoia, to be issued to the 'citizens and shopkeepers'.[26] The Papal States undertook a similar centralized purchase and distribution of a range of weapons, arquebuses among them, for purposes of civic defence.[27] Defence policy across the Italian states in this period encouraged training with firearms, and even those men who did not participate directly in the Italian Wars as soldiers are likely to have gained some familiarity with guns.

The risks of gun proliferation become even more apparent once military contracting, a central mechanism in the warfare of this period, is taken into account. The Italian Wars were fought between complex and shifting alliances of states and with extensive involvement of mercenary companies, a system of outsourcing that did not favour tight central oversight of weapons and their whereabouts. Demobilization was often chaotic, and soldiers left without wages owed, which contributed to a rise of banditry and other criminal activity.[28] Moreover, in the military context at least, some arquebusiers supplied their own guns, and it was common for them to receive a higher wage in return for purchasing their own match, powder and shot.[29] When an army dispersed, therefore, former soldiers were left with their guns, which might be abandoned, sold or retained in use for more or less legal purposes. Members of the early Florentine militia are known to have sold or gambled away their weapons (this, along with the purchase of such militia weapons, was subsequently banned).[30] These were Europe-wide problems: tapestries of the 1535 conquest of Tunis produced to Jan Cornelisz Vermeyen's designs show Imperial soldiers gambling for booty (Figure 5.2); similar rules against gambling and sale of arms are to be found in ordinances of the Grand-Duchy of Lithuania.[31] Regulations also set out what should happen to guns on an owner's death. In Brescia, for example, efforts were made to recover the arms of militiamen who died or otherwise left the service; these were to be deposited with the captain or rector of the city so that they should not be used for evil or sinister ends. The responsibility for ensuring the deposit lay with the *comune* to which the militiaman belonged. The wording of the ducal letter setting out this provision points to a contemporary perception that guns could quickly shift from a legitimate object in civic defence to something 'bad' or 'sinister'.[32] This was the challenge facing the Italian states: how to promote the use of this new technology for purposes of defence while preventing its abuse.

We will look at militia regulation in more detail in chapter 6, but for now let us consider just one case that illustrates the potential for guns to slip from legitimate militia uses to illegitimate ones. This concerns a soldier called Battista Magnani, who in 1577 was a member of the local guard in the Bologna countryside. Battista was a nephew (or other younger relative) of the parish priest and was alleged to have threatened another man (and perhaps also his wife Girolama) with an illegally short wheellock gun and a dagger. Numerous witnesses were questioned. Some of them had seen Battista (or perhaps another man) drop such a weapon while dancing at a party.[33] A number, however, had seen this small gun 'in the

FIGURE 5.2 Jan Cornelisz Vermeyen (design), Willem de Pannemaker (tapestry), *The Conquest of Tunis* (detail), 1548–54, Royal Palace, Madrid, Spain. © Paul Maeyaert. All rights reserved 2025 / Bridgeman Images.

guardroom', one witness specifying that it was on a table there.[34] There it would have been accessible to Battista but (by implication) also to other people. Asked about the armoury, Battista himself explained that wheellocks were indeed kept there; the presence of the (legal) guns was confirmed by another witness.[35] Battista claimed that two other men, Benedetto and his cousin Jacomo, had brought short wheellocks into the guardroom, but insisted that otherwise he had only ever long (legal) ones there. Some witnesses were prepared to confirm that Benedetto had a short wheellock in the guardroom, but others questioned the story that Battista had carried a short gun, one insisting that 'if Battista had carried it on his belt I think I would have seen it'.[36] Whoever was telling the truth here, several witnesses were in agreement that the militia store, which should in theory have been a secure and regulated place to deposit weapons, was also accommodating one or more illegal guns. Indeed, it is possible that the short guns discussed here were the equivalent of the later sawn-off shotgun, adapted from the legal militia weapons rather than specially produced. Multiply this incident by the hundreds of local militias across the Italian peninsula and it is easy to see the challenges of policing.

Contests and the hunt

If militia service provided one route to familiarity with firearms, so did the related rise of shooting contests. In 1506, Venice introduced a competition to encourage the development of handgun skills among the archers serving on its naval galleys.[37] To some extent, this followed practices already long established elsewhere in Europe. In the Holy Roman Empire shooting contests were principally sporting events, the priorities 'bourgeois pleasure and town glory' as local shooters sought to beat rivals from elsewhere.[38] Yet while the imperial contests facilitated diplomacy between towns and rulers as much as military preparedness, the Italian documents place the emphasis on the latter, and it is not evident that invitations to participate were issued to outsiders at all. Overseeing the Venetian contest was a committee of six senior city officials, including members of the heads of the Council of Ten, whose surviving records show how the contest worked in practice.[39] A set of rules from 1531 explains, for example, that the *schioppetto* was to be shot without a rest.[40] (Only in 1569 was the use of the large arquebus, *archibusone*, incorporated into the contest, alongside the even larger falconet.)[41] In earlier years, the prizes included lengths of costly fabric. In 1518, for example, the first prize was sixteen *braccia* of crimson (*cremesin*) damask, worth 32 ducats (a *braccio* is approximately an arm's length), along with 6 ducats in cash.[42] In 1520, the Republic of Lucca instituted a thrice-yearly shooting contest in order to ensure that its citizens were prepared for defence, the decree urging them to prefer 'death over the loss of dear liberty and the reduction of their fatherland to servitude'. Prizes included a silver cup worth almost 20 ducats, and six *braccia* of damask worth 5 ducats.[43] From 1541, the prizes in Venice were given exclusively in cash, with a total fund of 175 ducats (the lists suggest seven or eight prizes were usually awarded, making the average prize over 20 ducats).[44] Soldiers' pay could be less than 30 ducats a year,[45] so these were generous rewards and it is therefore not surprising to find evidence of cheating. In 1560, for example, one contestant tried to enrol in the Venetian contest twice under different names; he won the second prize, only for it to be confiscated and awarded to the person who had reported his misdeed.[46] Contests, however, were not only about financial incentives. They were also about establishing an acceptable social image for the firearm, just as the iconography of luxury weapons had done at court. The language of the Lucca decree, for example, offered a new and more positive discourse of the gun, as a weapon for the defence of liberty.

Hunting was another context in which guns became familiar not just for elites but more widely. At the start of this book, we encountered Vincenzo the cowherd, who in 1552 obtained a licence from the Bologna authorities to carry a gun for hunting, as well as the anonymous critic who complained of the wide use of wheellocks by herdsmen and shepherds. In the present day, a sharp distinction is often drawn between hunting and military uses of firearms, but Renaissance writers, including Machiavelli, treated the hunt as useful preparation for warfare in terms of the need for hunters to understand terrain and endure long days in the saddle.[47] A series of sketches by Jan van der Straet feature guns (Figure 5.3). These images of men hunting deer, ibex and geese were prepared for a series of engravings of hunts published in 1578 and then in an expanded version in 1594.[48] The figures shown are not obviously noble or aristocratic, confirming once again the wider social use of guns for game shooting. Indeed, in 1568 one man accused of a firearms offence, Giovanni Battista da Ravenna, excused his behaviour by claiming he was heading out of town to go fowling.[49] He kept a matchlock in the city of Bologna, in his master's kitchen. This may not have been illegal in itself (depending on his master's rank), but Giovanni Battista was accused of loading the gun and lighting the cord while still within the city limits. Questioned as to why, he insisted that he was planning to go and shoot some birds that he had seen, and that he had lit the cord only on his way out of the city. Later, he admitted that he had lit it in the garden.[50] His initial effort to defend himself relied on a shared understanding that countryside fowling was a legal and acceptable use of firearms.

FIGURE 5.3 Jan van der Straet / Giovanni Stradano, *Hunting Ducks*, 1578. Photo: Penta Springs via Alamy.

Guns in the criminal records

As guns became more common, so did their appearance in the criminal records. By 1600, firearms had become the predominant weapon used to commit homicide in Bologna.[51] That said, according to Colin Rose's study they still accounted for only a third of offences, and in interpreting the figures we should bear in mind the relative lethality of guns when compared with other weapons, and that a homicide attempted with a gun was probably more likely to be successful than one attempted with a knife (especially given that this was a society where everyone carried a knife and would have one available for self-defence). In fact, as Joe Tryner's research on the earlier Bologna records has confirmed, across the broad range of gun crime it is relatively rare to come across cases where guns were directly used in planned homicide.[52] More commonly, they were used to threaten: the ideal outcome for the highwayman who shouts 'Your money or your life!' is to ride away with the money without firing a shot. The limited use of firearms in homicide may also reflect a level of restraint on the part of gun users, who surely understood that the consequences of a shot hitting an individual were potentially very serious. Gun offences can be divided into two categories: those where a gun was introduced into an existing type of crime, and those arising specifically from the criminalization of certain types of weapon (which often arose from concerns about their use in the first category). To some extent, the policing of the latter sought to prevent the former.

Many of the historic investigations feature possession of illegal weapons and/or their use to intimidate. For example, in a Brescian case of 1554, Valerio Soncino and a gang of twenty men armed with various weapons, including arquebuses, scaled the walls of a property belonging to Valerio's brother Annibale, and 'with various threats' forced two of his *massari* (stewards) to bring them carts, which they loaded up with goods, stealing in addition cattle and a horse. The robbers got only five miles, however, before they were halted by the villagers of Pompiano, who had got wind of what was happening and sounded the alarm. In the ensuing violence, two men and a woman among the villagers were injured.[53] A case in the following decade also saw guns being used in a dispute between family members. In this case, Scylla Martinengo, a nephew of the complainant, had shut himself inside the family fortress 'with falconets and arquebuses' and was extracting money.[54] (The Martinengo were prominent pro-Venetian feudal lords with large landholdings in the area around Brescia.) A captain was sent up to investigate, but Scylla and his men (ten to fifteen of whom had arquebuses) threatened and laughed at him. The statements collected by the rectors include a particularly detailed account of how Scylla had used his gun to intimidate a man named Vincentio who he was trying to rob, describing how Scylla 'lowered his gun and put its mouth against [the witness's] stomach, and made as if to fire it'. Scylla, realizing that Vincentio really did not have any money to give him, used the gun to beat Vincentio on the chest and back. Subsequent witnesses described the mix of weapons in some detail including matchlock muskets, powder and shot plus wheellocks and also *schioppi da campo* (presumably handguns of a military type); one noted that some guns were worn on belts. In both these cases, the primary use of the gun was to frighten people into doing as they were told, not to kill them. Rather, they were worn by bandits as a means to threaten, in which capacity they proved effective. None of this is to say, however, that the proliferation of firearms did not create risks. The Bologna records describe a case in 1575 in which an elderly woman named Giovanna was killed by a man named Pierantonio Fangarezzo; Fangarezzo had been looking over an arquebus *all'acciarino* (meaning literally 'with a steel'; perhaps a snaphaunce, an early variety of flintlock) to see how it worked when it accidentally went off.[55]

A further group of people with an interest in firearm use were those engaged in, or considering, political violence. The distinction between criminal and political activity was not always straightforward. The system of exile in the Italian states—in which offenders were banned from their home territory and their property confiscated—often made banditry an attractive option (hence the overlap between the concepts in Italian, where the word *bandito* is also used for an exile).[56] Civic authorities were perhaps particularly keen to police weapon ownership when it posed a risk to the regime. The early years of Medici ducal rule in Florence (formally established in 1532) saw numerous confiscations of illegal weapons, often linked to charges of sedition: Girolamo di Filippo Bonciani and Francesco di Piero Serragli, for example, fell foul of the law after an arquebus was found under the straw of their former stable.[57] Some of these weapons were probably planted by city officials, but opponents of the regime might reasonably have decided to arm themselves in case of an uprising (for which there were several precedents).[58] Dissidents in Brescia also obtained weapons: testimony from 1539 shows how one alleged firearm offender, Antonio Maria, was accused of saying several times that he wished the Emperor and the king of France would unite against the Venetians and come and sack Brescia. Antonio Maria was further accused of 'continually keeping in his house exiled men of a bad sort, who go out killing on the streets'; among those he had taken in was a Julio da Spin [*sic*], exiled by the duke of Milan, and he had in his house around twenty-five firearms between muskets, arquebuses and *schioppi*.[59] While we typically hear about these cases from regime records, by reading between the lines it becomes evident that at least in some notionally criminal cases there might also be a political motivation for gun ownership. In chapter 7, we will see how guns came to be used in several prominent assassinations.

The contours of ownership

People in northern and central Italy owned guns for several reasons: to participate in militia service, for hunting and livestock control, and for criminal activity. Some of this ownership was purely practical: a gun would more reliably kill the fox eating your chickens than the alternatives. Some of it had political motivations, whether to defend the liberty of one's own fatherland or to oppose a tyrannical overlord. By the middle of the sixteenth century, a lot of people had guns. It is challenging to establish quite how many. Based on population estimates and numbers enrolled in the militia, Michael Mallett and John Hale estimated that one in fifteen rural households in Venetian territory would have had a firearm following the establishment of the militia in the later 1520s: 'the government', they observed, 'was calculatedly accepting a risk to public order'.[60] By 1540, the ruler of Modena required that one in three countryside households should have a gun.[61] These calculations concern militia service, however, and we know that other people may have bought guns second-hand from soldiers, or won them via gambling. Later in the century, larger cities had shops where guns might be mended and probably also purchased. A 1559 *bando* (proclamation) issued in Bologna in anticipation of riots required sellers of polearms and arquebuses to store their stock in such a way that it was not ready to use.[62] How common such sellers were is not clear from the available sources, nor are the retail prices, which in any case would be complicated by second-hand trading and the widespread practice of purchasing on credit. We have seen, however, that guns could be purchased wholesale for about 2 ducats, which represented less than a month's wage for even a poorly paid infantryman.[63] There is no reason to think that the anonymous author who complained of wheellock-toting shepherds was greatly exaggerating the extent of proliferation.

Historians have often turned to inventories to address questions of what people owned and, while the limits of these sources are well established, used with care they can still provide useful indications. This is especially true for elite households where inventories were taken for reasons other than post-mortem tax assessment (such as on handover of an administrative role from one holder to another). As we saw in chapter 4, ninety-one firearms were recorded in the wardrobe inventory of Duke Cosimo de' Medici for 1538/39, including four wheellock handguns made in the German style for use when travelling; an inventory for 1554 referred to senior Florentine courtiers and military officials using wheellocks.[64] A list of items for transport between Orsini properties dated the same year included eight arquebuses and four shields (*rotelle*) in a list primarily of carpets, storage chests and camp-beds: here guns were simply part of the furniture.[65] A 1569 inventory of the *mobili* (portable goods) of Giorgio Santacroce (member of a prominent family of Orsini vassals) included twenty arquebuses for use on horseback (*a cavallo*) and twenty described as *a braccia*, presumably for use on foot.[66] Reference to a vineyard suggests this list may refer to goods in a countryside or suburban property.

For wider society, however, the study of inventories can throw up as many questions as answers. The Refashioning the Renaissance project, led by Paula Hohti, investigated seven hundred inventories of artisans in a database covering the period 1550–1650 in the Italian centres of Florence, Venice and Siena along with the Danish town of Elsinore. However, the Italian sources included only two listings of guns, both in a single inventory, and both wheellocks. This is partly to be explained by differences between rural and urban firearm use: as we will see in chapter 6, much heavier restrictions applied within the cities than outside them, and the guns of militiamen were to be returned to central stores on their death.[67] The exceptions in the database are instructive about the types of people who were permitted to hold personal weapons within cities. The wheellocks (one long, one *da braccio*), were listed in the 1597 inventory of Paolo Rimbombi, a *trombetto* of Siena.[68] A *trombetto* (more commonly *trombetta*) was a trumpet player, but the term often implies a role analogous to the English 'trumpet' of the period, who might play a fanfare prior to making a public announcement.[69] If Rimbombi had to travel around the Sienese countryside to do so, then he may have been permitted to carry a gun. In contrast, the 1572 inventory of Antonio del Grasso, a Florentine *bombardiere*, makes no mention of firearms, only of a (silver) dagger and sword.[70] Given his profession and the date that might seem surprising, but in light of the regulations it makes sense that any guns he used would be have been returned to central stores prior to preparation of the inventory (or else kept outside city bounds or indeed concealed from the notaries). The project team identified significantly more guns and accessories in personal ownership in their comparative study of Elsinore, but we should not assume that the Italian artisans were less familiar with firearms than their Danish counterparts, rather that the legal framework in which they used guns was different.

Certain moments in wartime no doubt accelerated access to firearms. We have already noted the artisans issued with guns in the course of the 1529–30 Siege of Florence. Records of the defensive provisions in Ferrara of 1552 (when the city was likewise preparing for a possible siege) show that out of 520 households equipped with armour, pikes, arquebuses and morions, more than half (270) had a gun. Included in this list are eighteen female heads of household, of whom six had guns.[71] This raises the question of how many women were gun owners more generally. The answer seems to be rather few. Sieges were emergency situations in which gender norms might temporarily be suspended, such as when the women of Siena took responsibility for repairs to the city walls.[72] Beyond the wartime context, Tryner's study of court records in Bologna turned up only a single example of a woman with a firearm, and nothing in my wider survey of source material has suggested that was untypical.[73] A clarification that gun laws applied without distinction of sex was added to the Bologna *bandi* from 1580.[74] It is impossible to say whether this change was prompted by an individual incident or a wider perception that it was better to be clear that the laws applied to both men and women, but Catterino Zen's 1553 observation, previously encountered in chapter 2, that 'even the women' of the gun-producing area of Valtrompia carried wheellocks certainly implied that women's use of firearms was particularly undesirable.[75] Given the context, this may also have been a pejorative comment on Protestant ideas about gender. Either way, however, the sources provide many more examples of women depicted on guns than they do of female gun users. Lois Schwoerer likewise found limited evidence in the English sources for women's

FIGURE 5.4 Unknown artist, *Kenau Simonsdochter Hasselaer*, c. 1590–1609. Oil on panel, 34.8 × 26.5 cm. Rijksmuseum, SK-A-502. Public domain.

handgun use in the sixteenth century, though somewhat more in the seventeenth and eighteenth.[76] On the other hand, certain continental cultures may have been more open, particularly that of the Low Countries, where we find a notable portrait of Kenau Hasselaer, a shipbuilder mythologized for her defence of Haarlem during the siege of 1573 (Figure 5.4), showing her with multiple weapons including a wheellock gun. That should not come as a surprise, however, given the well-established differences in cultures of gender and honour between northern and southern Europe: while women were not full members of the firearms guilds of northern Europe, they participated in their social events.[77]

In contrast, however, guns did come to be associated with masculinity. In relation to Florence, Nicholas Scott Baker has argued that the siege of 1529–30 was a particular turning point for elite identity, redefining notions of masculinity from those associated with practices of civic governance to something more 'rugged, militaristic'.[78] This was particularly apparent in changing clothing styles, to which the bearing of new types of arms might be added. Victoria Bartels, focusing on Florence after 1537, identifies guns as one factor in shifting performances of masculinity, noting that they increased the danger attached to 'routine bouts of aggression' that had traditionally been employed to settle disputes between men.[79] In the context of the Italian Wars (and specific local conflicts within them), guns had a role to play in changing ideas about gender both in general and with reference to images of rulership. This is confirmed in the account of Benvenuto Cellini, whose self-aggrandizing narrative of his participation in the Italian Wars includes a comment he made to a companion, Alessandro del Bene, to persuade him to stick around and shoot back at Bourbon's army during the Sack of Rome: 'Now you've brought me here, we must show that we're men.'[80]

Conclusion

Over the course of the sixteenth century, familiarity with firearms grew. It was increasingly expected that men would be able to use a gun: the expansion of militias was a key factor here alongside direct soldiering experience. The acquisition of skills in shooting was positively encouraged through civic contests with generous prizes. Rhetoric about the importance of guns for civic defence and for the maintenance of honour and liberty offered a counterweight to more negative images of gunpowder weapons as diabolical, unchivalrous or impious. Guns were useful for protection of livestock from predators as well as for hunting more generally. In turn, their ready availability also made them accessible for use by bandits, during vendettas and indeed for political violence. However, their lethality shifted the dynamics of violent encounters: so effective were guns as a threat that in some cases they may have prompted victims to give up without a fight. On the other hand, with increasing proliferation came an increased risk of both accidents and casual use in interpersonal violence. People were at least somewhat different with guns in their hands. Given the threat to social order, it is no surprise that governments sought to regulate their use, especially in urban environments. The firearm revolution had implications not only for the military but also for the disciplining state.

CHAPTER SIX

Gun Control

IN DECEMBER 1551, THE rectors of Brescia wrote to the Venetian authorities noting that a local gentleman hosting a party had put guards on the doors 'to avoid any disorder' but that there were so many weapons at the event that in light of the 'many discords' in the city 'it was almost a miracle . . . no harm was done'.[1] Easy access to arms was a risk, and guns even more so given their lethality. The rectors, however, were locked in an argument with their superiors in Venice about how tightly to enforce the law on arms-bearing in general and guns in particular. Increased use of guns by the militia had brought with it wide access to firearms, and that posed risks to social order. There were increasing demands to carry guns in self-defence. These issues could become highly politicized, as we saw in the case of Florence in chapter 5. There, during the establishment of the Medici duchy in the 1530s, the authorities clamped down on firearm possession in the city (and may have planted guns to justify arresting political opponents). Unsurprisingly, questions around arms-bearing became wrapped up in wider debates about infringement of liberty.[2] Over the course of the sixteenth century, civic authorities developed extensive systems of regulation to control who could carry guns, where, when, and in what circumstances.

Firearm owners were not only from the higher social ranks. Among the men who obtained gun permits in Bologna (primarily for shooting in the countryside) were a saddler, a shepherd, a tailor, and a builder.[3] People at all levels of society had to navigate the gun control legislation. Think back to Vincenzo, the cowherd, who obtained his gun licence in Bologna for the purpose of hunting. The 1552 register that recorded his request illuminates the questions considered in the process of issuing a permit to carry any type of weapon. Would it be carried in the city or the countryside? During the day or at night? Would it be concealed? The register and associated legislation show that self-defence was already a motivation for requests: Carolo Fontana, a surgeon, was allowed to carry unspecified arms both day and night in the city for that purpose.[4] Certain categories of people were automatically entitled to carry arms and therefore would not be expected to appear in these registers: they included city residents and their agents (*fattori*) who were travelling with weapons between their city and country residences; members of the government and their households (with some restrictions on numbers of servants and the time of day they could go out armed); senior guild and college officials and their servants; and so on. Broadly speaking, the list of exemptions encompassed the senior officials of the city administration as well as those with the rank of knight or above.[5] Sometimes, however, permission to carry the most heavily regulated type of gun, the small wheellock, was not extended even to members of the nobility.

This chapter explores how states sought to regulate firearms, considering first the general development of gun laws before turning to three challenges faced by advocates of tighter regulation: the existence of historic privileges that allowed men to bear arms, shifting military practice that favoured wider use of firearms including small wheellocks, and finally the emergence of arguments about the use of guns in self-defence. These became increasingly prominent in the second half of the sixteenth century and reflect the weakness of the state in asserting a monopoly of violence as handguns became ever more common.

General regulation

While the Italian states encouraged familiarity with firearms for purposes of civic defence, they also regulated use of handguns, especially the small wheellock guns known as *archibusetti*. These could be concealed on the person, which made them the focus of particular concern, and restrictions generally focused on weapons of less than three palms (approximately fifty-seven centimetres). Such guns were not routinely used on the battlefield except by elite cavalry units (the mechanism was expensive and less reliable than a simple match), so regulation of their use ought not to have had a significant impact on civic defence. This was one way to resolve the conflict between defence and social order: the anonymous author who lamented cowherds' access to wheellocks suggested that if armies relied exclusively on matchlocks, then wheellocks might be outlawed altogether.[6] In practice, however, most states allowed exemptions: those permitted to carry wheellocks included elite bodyguards, and sometimes (though not invariably) individuals of high rank. Over the course of the sixteenth century, however, wheellocks proliferated to the point that they became the weapon of choice for both bandits and travellers seeking protection from attack. Besides their advantages for night-time or twilight hunting, in the context of personal or household defence they were more convenient than keeping a match constantly lit, though, as we saw in chapter 4 and will see further below, this view was open to question.

The first specific European ban on wheellocks came in 1518, when the Holy Roman Emperor Maximilian prohibited them in Habsburg territory.[7] Such bans eventually extended to cover most of western Europe, although the authorities struggled to make them stick.[8] In a decree of 1513, the duke of Ferrara had already ruled that no one 'of whatever condition', including those in his own pay, should go about in the city at night with *schioppetti* without his express permission (the ban also extended to crossbows and polearms). A subsequent decree of 1522, however, shows that the ban had been ineffective. 'Under the pretext of the tumults and movements of war that there have been in these past months,' it read, 'some people have taken licence to carry the said prohibited weapons and to go about at night as they please, which, for the aforesaid reasons, our Illustrious Lord has in part tolerated, because that is what the condition of the times required.' This is an early example of a ruler acknowledging, at least implicitly, that there were occasions when carrying a gun in self-defence was understandable. Once the tumults were past, however, the ban was reiterated, and to it was added a specific reference to guns *da fuoco morto* (dead-fire), presumably wheellocks.[9] Bologna prohibited the carrying of *schioppetti* in the same year.[10] Subsequent reiteration of the bans in both cities, however, suggests that people continued to ignore them.[11]

Small wheellocks were targeted across the Italian peninsula. In 1532, Venice banned both the production and the use of *schioppi* 'so short that they could be hidden beneath clothing', citing the recent murder of Antonio Bendenuzo, a captain, as motivation for the new restriction; while there is evidence for the arrest and imprisonment of those who infringed the law, the ban nonetheless had to be reiterated in 1545.[12] (Concerns about concealed carry were not restricted to firearms: the 1558 statutes for the Duchy of Castro and Ronciglione, north of Rome, imposed a double penalty for concealed carry of any illicit weapon within Castro and the other towns of the duchy.[13]) An explicit ban on small wheellocks was in place in Sicily by 1530, again with subsequent reiterations.[14] Milan, which from 1535 was ruled by Spanish governors, had prohibited wheellocks by 1538.[15] Duke Cosimo de' Medici banned small guns of all types (less than one-and-a-half *braccia* or 87 cm long) in his Tuscan territories in 1547: this specific edict followed the inclusion of guns in a wider-ranging ban on arms in 1539.[16] In Bologna the explicit ban on short wheellocks was reiterated in 1550.[17] Many restrictions on small wheellock weapons were enacted, therefore, after the end of the major land wars between France and the Holy Roman Empire in Italy, which had concluded with the peace treaties of Barcelona and Cambrai in 1529. Like the Ferrara ruling, they reflected a perception that once conflict had calmed, there was no excuse for the carrying of such weapons.

This was not entirely a post-conflict environment, however: there were ongoing threats of smaller-scale warfare, notably in the north-west of the peninsula near the border with France and Savoy. The efforts of Pope Paul III's family to carve out for themselves a new Duchy of Parma and Piacenza, as well as the conflict over Siena, also led to war in the 1550s (involving once again French and Spanish forces). While the larger western powers settled their differences with the Treaty of Cateau-Cambrésis (1559), albeit leaving some specific issues unresolved, the Fourth Venetian-Ottoman War followed in 1570–73. This was primarily a naval conflict, but Ottoman incursions onto the Italian peninsula were familiar enough, especially along the coasts and islands of the Kingdom of Naples.[18] It is not surprising, therefore, that the firearm restrictions of the 1530s and 1540s targeted gun users and not manufacturers or merchants: the geopolitical situation favoured the maintenance of an industry that could readily provide the Italian states with further weapons at a rapid pace should they be required. Only occasionally does one find cases in which manufacturers fell foul of the law: for example, in January 1549, when three masters of arquebuses were arrested in Brescia after short wheellocks were found in one of their houses.[19] Targeting of producers was rare, however, perhaps because rival rulers were still competing for their technical expertise and heavy-handed policing would have risked encouraging emigration.

Enforcement of gun control legislation took a variety of forms. If found, illegal weapons might be seized: some apparently made their way into the Medici wardrobe, where in the 1560s were to be found a small wheellock arquebus, 'which they say was found in the baggage of some Germans', and two large ones, allegedly 'left in Calcione by certain bandits'.[20] (Given the context these may have been exiles rather than robbers.) In addition to straightforward prohibitions, the authorities also tried more indirect measures, such as those in Bologna requiring surgeons and barbers to report cases of wounding to the authorities within twelve hours (if in the city) or twenty-four hours (if in the countryside).[21] A similar system operated in Rome.[22] The problems persisted, however. In 1560, a decree in Ferrara reiterated the ban on small wheellocks, with the observation that 'so much are they multiplied in this dominion . . . and so licentiously are they carried by everyone, not without the danger of many ill effects that might be born from such arms, which reasonably ought to displease everyone'. In 1573, the ban was supplemented by a stop-and-search provision allowing the duke's officials to search houses, shops and people, specifically targeted at small arquebuses 'that could be carried in breeches or sleeves or otherwise concealed and secret'.[23]

All that said, the law left wide scope for ownership and use of firearms beyond the problematic small wheellock. Longer-barrelled wheellocks were acceptable, as were most types of matchlock, albeit with some limits on where they could be carried. Besides regulation by type of gun, there was also regulation by place, made possible by the clear delineation that city walls provided. These limits largely mirrored earlier city regulations on the bearing of arms that restricted which weapons could be carried within the walls and limited the wearing of swords (for example) to higher-ranking men. The firearms licences issued in Bologna typically provided for gun use either in the countryside (by implication, not in the city) or specifically for hunting. When guns were permitted, the licences often reiterated the rule (also found in general gubernatorial proclamations known as *bandi*) that guns must not be carried in markets or churches, or at public festivals.[24] Different rules could apply by day and by night, and for people of different rank. In all these cases, the legislation was refined in light of experience. In 1555, for example, it was observed in Bologna that some people had found a means of carrying concealed small matchlock arquebuses: the loophole was subsequently closed.[25] The fact that such modification had been achieved, however, points to another significant issue: the development of skills in adapting weapons. These may not have been particularly high-tech in some cases (the equivalent of sawing off a shotgun); moreover, the wind-up technology of a wheellock was similar to that of a clock, and clocks were becoming more common.[26] (This may explain the production of the powder-flask with a clock on it now in the Luigi Marzoli museum,[27] as well as the example shown in Figure 4.2.) Peter Peck, known for his luxury wheellocks, was documented in Munich in 1540 as a watchmaker.[28] Moreover, as we saw in chapter 3, while military authorities took care to supervise inventories, items still went missing. There were thus multiple means by which it was possible to obtain firearms that infringed the rules.

Privileges, social status, and gun licensing

A significant barrier to gun control was the existence of longstanding privileges that granted a range of men in sixteenth-century Europe permission to bear arms. These included not only individuals of high rank, but also in some cases their servants, as well as city officials and members of the militia. At this stage, such ideas were almost always expressed as 'privileges': the discourse of 'rights' came later, in the seventeenth century. These privileges were jealously guarded by those who enjoyed them. In February 1551, the doge of Venice wrote to the rectors of Brescia with orders to revoke certain penalties imposed on those carrying arms because they infringed privileges enshrined in law.[29] (This decision was made in the rather delicate context of Brescian subjection to Venice: the subject city's old patrician families continued to wield significant power and autonomy in respect of their activities, which the rectors could not entirely control.) Brescia was not the only city to face problems establishing the boundaries of privileges: in 1544 a proclamation from the legate in Bologna set out numerous rules regarding exceptions: for example, that those authorized to carry arms in their capacity as servants to the ruling Council of Forty (each senator was entitled to nominate three such servants) could only do so after the evening bell if they were in the company of their employer.[30] A Brescian register of 1551 shows that gentlemen in the retinue of the duke of Urbino, captain-general of the Venetian army, were allowed to keep small wheellock guns (otherwise prohibited) at home and carry them when in the duke's company; the cavalry were also permitted to carry them when on official duties.[31] Sometimes, concern was such that bans were quite extensive. As we saw in chapter 4, one of the first recorded people to buy a gun 'fired with a stone', almost certainly a wheellock, in 1507, was Cardinal Ippolito d'Este, a member of the ruling family of Ferrara.[32] Yet while the cardinal and his relatives might collect such fine weapons, a prohibition on wheellocks issued in Ferrara in 1551 spelt out that the ban applied to everyone of whatever rank, whether subject or foreigner, 'including also those of the court of His Excellency and the Most Illustrious Madama and the Lord Prince', unless they had express licence from the duke.[33] The desirability of the new weapons combined with their risk was creating significant tensions.

To some extent, privileges could be managed by licensing and registration. In 1546, for example, Giovanni Battista Porcellaga, whom we met in chapter 3, was issued with a licence to carry arms that covered Brescia and Venice as well as other cities of the Venetian dominion. His licence encompassed not only himself but also his sons and one servant for each of them; however, these could only be household servants receiving a salary and board: he could not simply delegate this right to anyone. Moreover, when visiting Venice, the names of the authorized servants (the licence itself did not name individuals) had to be provided to the Office of the Night in Venice, or to the Cancelleria in Brescia or another subject city.[34] This kept a check on the misuse of licences by persons not clearly linked to a noble household. Porcellaga's licence also introduces some of the expectations of those authorized to bear arms. He had, according to the licence, shown 'faith towards our state and good qualities'. This was a standard formula. A 1559 licence for members of the Martinengo, another elite family of Brescia, likewise referred to 'fede et fidelissime operationi verso il stato nostro' (faith and most loyal operations towards our state).[35] While these licences are somewhat formulaic, they do show subtle differences that indicate the variety of qualities for which a licence might be granted. Lelio Palazzo, granted a licence in October 1553, was simply a 'Brescian gentleman'.[36] Girolamo Luzzago, 'fidelissimo et benemerito del stato nostro' (most loyal and well-deserving of our state), was allowed a licence for himself and two servants.[37] The rector Catterino Zen, whom we have encountered in chapters 2 and 5, noted in 1553 that the privilege of arms-bearing was widely extended in Brescia, more so than elsewhere.[38] Certainly there are numerous examples of requests for arms licences from petitioners citing a range of justifications, including individual rank as knights or counts, historic loyalty to Venice and service in its military, and virtuous living.[39] Where a family had historically opposed the Venetians, more generic language around honour might be used instead.[40] Noblemen were probably most concerned about the privilege of wearing a sword (the licences are not specific), but by this time it was increasingly common when travelling to employ firearms for personal security, so it is likely that in practice the permits covered a variety of different weapons.

Restrictions on the privilege of arms-bearing could create tensions between state authorities, the nobility and courtiers, but in practice relaxed policing meant well-connected individuals often faced few consequences for breaches of the gun laws. When in 1548 Cardinal Doria's son was caught with a small wheellock in his bag (which infringed Venetian gun laws), he was initially arrested but subsequently freed 'out of respect for his father', though his gun was confiscated.[41] The Doria were the de facto ruling family of Genoa, and there was clearly an expectation on the part of higher-status individuals in those states that imposed bans that they should enjoy more liberal treatment. Nor was this restricted to the nobility: it might also extend to professional men. In 1538, the Brescian rectors sent on to the heads of the Council of Ten in Venice a confiscated handgun which discharged 'artifically and without fire'.[42] However, they also accepted the word of the man arrested with it, Zuanagnolo de' Taieti, that it was not his weapon. Zuanagnolo held the title of doctor and had given a convincing enough explanation for the provenance of the weapon, whose original owner was now deceased.

Men of lower ranks caught with illegal guns could also benefit from patrons' lobbying. In 1540, the Cardinal d'Ivrea, Bonifacio Ferrero, requested that Cosimo de' Medici show clemency towards a French student detained for carrying an arquebus.[43] This student evidently did not learn his lesson, because he appears to have been arrested a second time five years later, once again relying on an influential patron, in this case Cardinal Benedetto Accolti, to petition for his release.[44] In 1551, the Sienese authorities wrote to Cosimo to appeal in the case of two citizens, Mario Ottorenghi and his servant boy Simone, who had been caught in Florentine territory in possession of a small wheellock arquebus and had respectively been fined 25 scudi and sentenced to two drops on the *strappado*.[45] (The *strappado* was a common form of punishment in which the subject's hands were tied behind his back and he was suspended from his tied arms and, in this case, dropped with a jerk.) Lobbying likewise took place in Venice, where in 1547 the English ambassador obtained the release from prison of a Zuan Antonio Grumo, who had brought his small wheellock from England; imperial ambassadors secured similar concessions.[46] This lobbying echoes much longer-standing practice whereby diplomats might lobby for fellow nationals who wished to carry arms.[47] It shows, moreover, that in practice gun control laws were flexibly applied.

For those lower down the social ladder, policing was often much stricter. A Venetian case from 1530, for example, shows how a young man who had obtained a position in the city guard as a pretext for access to weapons—and then misused them—was sacked and banned from carrying arms.[48] The weakness of the state, however, is apparent in the less than effective implementation of the licensing system through which wider layers of society might gain access to guns. In Bologna, for example, the militia was exempt, which left a large number of weapons free for use in the countryside.[49] A city *bando* (proclamation) of 1544 shows that staff of the legate and vice-legate had been issuing arms licences without proper authorization. The officials in question were denounced as having 'fled the light and passed into the darkness', and those who had obtained such licences were encouraged to report the issuers, on the promise of anonymity and a reward of 25 scudi. (A scudo was worth about 6 per cent less than a ducat, so this was around two-thirds of a soldier's annual wage.) A subsequent proclamation, issued in 1552, revoked all previous licences conceded 'in writing or orally', suggesting that the process of licensing had been somewhat informal. The problem did not go away, because in 1563 a further *bando* referred to 'many insolent people who dare to carry both arms and armour without licence from His Most Reverend Lordship [the legate], and many others who use the licences conceded in ill part, from which arise many troubles and scandals'.[50] If, however, the licence book for 1552 reflects ongoing practice, relicensing would have entailed costs for gun users and provided income for the authorities, which may have been an incentive to exaggerate the problems. Licences generally cost 5 soldi (arquebuses themselves cost in excess of a ducat—that is, almost thirty times the cost of the licence) and brought in a total income of 107 lire in September 1552, 123 lire and 15 soldi in October, and 143 lire and five soldi in November, suggesting that the total annual income might be about 1,500 lire (this equated to about 200 ducats, which was by no means a fortune but nor was it a negligible amount).[51] There was also, of course, a risk of forgery, though an individual accused of making a false licence in 1576 was determined by the authorities to have been duped by others and was therefore ruled not guilty.[52] One man arrested in 1577 for possession of wheellock guns claimed in his defence that his cousin had said he would 'put him on the list' of those permitted to carry weapons.[53] Without consistent enforcement and written records, the licensing system struggled.

The problems were exacerbated by the widespread practice of tipping and gratuities in dealing with public officials, who relied on such payments to supplement their income. As we saw in chapter 3, in 1551, at a point when the Gambara were about to make several requests for licences to bear arms, the family put on a dinner for the podestà's son and 'some gentlemen'.[54] Indeed, there were also examples of more direct incentives for the authorities: as well as a payment of one ducat to the podestà's chancellor for a licence to bear arms, the Porcellaga accounts refer to gifts to local judges.[55] While in a present-day context that would undoubtedly be regarded as bribery and corruption, in the sixteenth century the lines were not yet so firmly drawn.[56] Between existing privileges, individuals of high rank exercising power or applying money to secure more favourable treatment for themselves or their associates, and a licensing system that was easily corrupted, there were multiple reasons why the authorities struggled to implement gun control.

Soldiers, militiamen, and regulation

Shifting military technologies, not least the increasing use of wheellocks by soldiers, also posed a challenge to the regulation of guns. In 1552, in light of ongoing problems with arquebusiers from the militia carrying weapons in Brescia, the doge of Venice conceded that while their privilege to carry arms should be observed, that privilege did not allow them to do 'that which should not be convenient', and the rectors might act accordingly, a political fudge that seems unlikely to have pleased anyone.[57] By the second half of the century, the arquebus had become not just a weapon for infantry but increasingly for cavalry. Records from Ferrara in 1557 show payments to fifty cavalry arquebusiers who formed the ducal guard in the city of Modena.[58] Given the difficulty of reloading a matchlock on horseback, we can fairly assume they used smaller wheellocks, and as such use became more common it challenged a legislative framework that focused on the dangers of such weapons. A 1578 document from the Duchy of Castro (a feud of the Farnese family north of Rome) shows how the *colonello* responsible for the militia wanted to waive the rules on minimum arquebus length for his cavalrymen but the ducal *auditore* disagreed that he had discretion to do so.[59] It is not clear how this conflict was eventually resolved, but it illustrates the growing tension between military considerations and gun control.

That tension is particularly apparent in a case of 1560 that brought Paolo Correr, podestà of Brescia, into dispute with Sforza Pallavicino, governor-general of Venetian forces on the mainland.[60] In late November that year, a rider was stopped and his small wheellock gun confiscated by the vice-constable of Brescia in line with an outright ban on such weapons enacted in 1558. Pallavicino, however, protested that the arrest contravened several clauses of his *condotta* (contract) with Venice. The first of these concerned jurisdiction over soldiers, which in routine cases fell to Pallavicino; the second stated he was responsible for disciplining his men in all cases except 'atrocities', and that even if this were such an exception it was a matter for the captain, not the podestà;[61] and a third specified that, for the duration of his *condotta*, 'his men of war should have free transit . . . through all the cities, lands and places of the Most Illustrious Dominion with their horses, arms, harness, baggage and other similar things pertaining to their sole use without any payment of duty (*datio*), salt tax or road tolls'. There was genuine ambiguity here: did the unspecified 'arms' in the *condotta* include the otherwise prohibited wheellocks, or did they not? A letter to the rectors some months later, in June 1561, suggests the Venetian authorities came down on Pallavicino's side. Without making specific mention of wheellocks, the doge highlighted the existing privileges of soldiers, stated that it was 'proper to their profession' for them to carry arms, and said they should not be 'molested' for doing so.[62] Military considerations had evidently prevailed.

The existence of separate church courts created further difficulties. In 1552, Catterino Zen, podestà of Brescia, wrote to the Venetian authorities asking for advice in the case of a priest who had 'dared to carry one of those banned *schioppetti*'.[63] The cardinal of Brescia had claimed that jurisdiction in the case should fall to the ecclesiastical courts, but Zen worried that this could create an unwelcome loophole in the law.

If regulating private military companies (and rogue priests) was a challenge, the militia should have been more straightforward.

Unlike military contractors, they were directly managed by the state and in many cases covered by similar rules to private citizens. In 1568, members of the Farnese militia were still banned from carrying wheellocks 'without licence from the *commissario*' on pain of death.[64] In 1594, the militiamen of Castro and Ronciglione were banned from carrying both staff weapons and firearms within one hundred *braccia* (around 60 metres) of a place where a party or fair (*festa*) was going on, on pain of three drops on the *strappado* and expulsion from the militia; they were also banned from carrying staff weapons, firearms and shot within city and town walls (or village ditches) at night, during which time swords were also required to be sheathed. Soldiers were also forbidden from going about at night in groups of more than three, a provision that made it easier to identify larger gangs of bandits.[65] Yet even while these regulations stayed in place, there were shifts in the language of militia ordinances. The 1594 militia ordinances for the Duchy of Parma and Piacenza explained that 'each soldier is obliged to bear those arms ordered by his Captain', to keep them well in order, and not to lend them out without permission. Moreover, if they went out to work in the countryside any distance greater than an arquebus shot from their house, they should take the gun and its accessories (or a staff weapon) with them. Soldiers, the ordinances concluded, should 'hold their arms most dear, as the most beautiful and honourable thing, that they have'.[66] The affectionate language towards firearms that we first encountered in the letters of dukes was now expected, indeed required, from country militiamen. The higher ranks of the militia (captains, lieutenants and standard-bearers) were further permitted to carry pistols everywhere except in the cities of Parma and Piacenza, and even there in cases where they were passing through on ducal service.[67] Similar permission was extended to cavalrymen, whose pistols were required, however, to have a minimum barrel length of six *onze* (half a *braccia*), and who had to deposit their pistols at the gates when transiting the cities.[68] Here, an argument similar to the 1578 debate in Castro was resolved with at least some concession to the cavalry. This wider use of small guns was the backdrop against which, in Bologna, firearms would replace knives as the main weapon in homicide, which was noted in chapter 5.

Arms-bearing for self-defence

A third challenge for gun control was posed by the increasing demand to carry arms for personal protection. By the later 1550s, concessions around arms-bearing for self-defence, at least outside the city gates, were being made to a wider section of the population than the traditional arms-bearing nobility and officeholders. These concessions become apparent in the preparations in Bologna for the *sede vacante* of 1559—that is, the period between the death of one pope and the election of another, frequently an occasion for social disorder. As was usual during such interregnums, the governor of Bologna issued a *bando* 'for the quiet of the city'. As we saw in chapter 5, this required sellers of polearms and arquebuses to store their stock in such a way that it was not ready to use, a requirement presumably aimed at mitigating the risk that a rioting crowd might seize weapons from local shops. The proclamation renewed the ban on wheellock guns and their production, offering a reward of 25 gold scudi plus a guarantee of anonymity for anyone reporting the possession or carrying of wheellocks. Repairers of wheellocks and their customers were likewise encouraged to report one another, for the same reward. However, the same *bando* also makes apparent that many people owned shorter firearms, because the governor now ordered that while small arquebuses (under three palms or eighteen *onze* long), whether matchlock, wheellock or snaphaunce, must not be carried by anyone within the city, they might be carried (unloaded and openly) by those travelling on horseback or in carriages between Bologna and a villa or other place outside the city. Open carry was also permitted in the city for purposes of repairing small arquebuses or for other good reason, but here again guns must be unloaded and uncovered. The penalty for breaking these laws was death.[69] The acceptance that guns could be taken for repair acknowledges that they might legitimately be used at least outside the city for defence of person or property.

Ideas about arms-bearing for self-defence became increasingly important in the middle years of the sixteenth century. A request from Camillo and Giulio Martinengo of Brescia, made in 1551, argued they should be entitled to bear arms for 'the security of their lives'.[70] As with Pallavicino's soldiers, however, there was a balance to be found when it came to policing the use of personal weapons. The Venetian Council of Ten was concerned at the abuse of arms-bearing privileges and wrote to the rectors in 1550 with a proclamation against those with licences who would draw their weapons 'without being provoked by others, and not for their personal defence'. This document provides a useful gloss as to contemporary understandings of how the privilege of bearing arms functioned: it was a defensive privilege, not an offensive one. Offenders would face the penalty of exile for three years and the withdrawal of their licence, or, if a servant, a three-year term on the galleys, and would in addition be required to pay 300 scudi (or 100 if a servant) to the accuser. A servant unable to pay the fine would have his term on the galleys extended.[71] It appears, however, that the rectors interpreted the Council's order rather more enthusiastically than the Council had intended, because in February 1551 it wrote again concerning a number of individuals who had been punished, including by exile or imprisonment, for infringing the arms regulations, leading to 'grave resentment'.[72] This exchange of correspondence illustrates the challenges for Venice and its officials of balancing on the one hand the need to maintain social order via regulation of martial culture, and on the other the need to avoid any sense that the authorities were impeding the exercise of properly granted privileges.

It is notable, moreover, that at times the climate of violence in and around Brescia forced the representatives of the Venetian state into conceding licences not so much on the basis of noble privilege but for self-defence. In January 1556, four members of the Patroni family (who do not appear on the lists of city officeholders) requested permission to bear arms, on the grounds that they had received threats from enemies. They noted that many other Brescian citizens had been conceded licences.[73] In other words, this was not merely a matter of honour for individuals or families, but of practical self-defence. The question of self-defence was also apparent in a request from members of the Da Monte family for a licence to bear arms. Forwarding the request to Venice, the rectors explained that the men in question were in their fifties, living quietly, but for unspecified reasons which were 'not their fault', they found themselves 'in no small suspicion and danger'. They were therefore requesting the right to carry weapons. The rectors had enquired with 'trustworthy people', confirmed that they lived quietly, that their ancestors had 'many merits', and that the Da Monte would not be making this request if not for their personal security.[74] Florentine sources similarly hint at self-defence as a growing motivation for arms-bearing, as in the case of the merchant Giovanni Caccini, on whose behalf Duke Cosimo de' Medici requested in 1559 that the viceroy of Sicily allow Caccini and his servants to carry wheellocks while travelling, in light of the risk of attack in that realm by 'ill-living persons'.[75] Self-defence is also the most likely reason why a homicide victim named Lorenzo, whose story we know from a criminal investigation of 1568, was carrying a gun. Lorenzo was murdered by a gang of four men (known enemies of his family) while on his way to harvest grapes in the Bologna countryside, not one of the agricultural tasks in which a firearm would typically be useful. The most likely explanation for its presence is that he hoped it would deter the enemies or be useful in defence.[76] Together these cases show that the decision to carry arms may often have been motivated by worries about personal security, rather than any great enthusiasm for guns. Self-defence, moreover, was not only a matter for the wealthy, or those of higher rank.

Travel was one aspect of everyday life in which the possession of guns was increasingly accepted as a sensible precaution against attack by bandits. Benvenuto Cellini took a wheellock on a journey from Rome to Naples in 1532.[77] While travelling between Rome and Virola in 1575, Cesare, a servant of Niccolò Gambara (of the Brescian clan), purchased gunpowder for a *schioppo*, and had an arquebus fixed in Ravenna, a city like Bologna within the territory of the Papal States.[78] A traveller arrested outside Bologna the same year for infringing the gun laws explained to his interrogators that he had the weapon for security while travelling to that city from Rome.[79] This was a plausible explanation because a range of cities allowed those arriving to deposit guns at the gates and collect them on departure: Bologna, Florence, Perugia and Rome had such systems.[80] Moreover, from 1551 foreigners in transit through Ferrarese territory could, at the discretion of the podestà (the senior judicial official), be exempted from the general ban on small wheellock firearms.[81] This was one way that wheellock weapons became normalized.

That said, there was already criticism of the idea that guns were useful defensive weapons in emergency situations. As we saw in chapter 4, Michel de Montaigne questioned whether they were effective in battle. The anonymous critic of firearms who complained of gun-toting cowherds pointed out that attackers would make sure their targets would not be able to use their guns, adding that it was 'hard to find an example of anyone who has been attacked with an arquebus and even if uninjured or unimpeded has still shot back at the enemy'.[82] On the other side of the argument were men like Catterino Zen, whose 1553 report on his posting discussed the widespread carrying of arms in Brescia. Zen believed that it was 'better to let them carry arms than not' because this reduced the number of homicides in the territory. During his fifteen-and-a-half-month tenure, there had been ten deaths in the city and fifty-two outside it, nine of out ten by arquebus, an improvement on previous numbers (which, however, he did not specify).[83] In a counterpoint to complaints about gun ownership, Zen saw generalized arms-bearing as a deterrent. He was not the only one: by 1580, the authorities in Perugia decided to permit the use of wheellocks 'of proper size' for self-defence.[84]

Conclusion

In the post-conflict environment, the enforcement of gun control laws depended on negotiation with gun owners. It ought not to come as a surprise that in the aftermath of the Italian Wars (1494–1559), the conflict in which handguns first came to prominence as a significant military technology, aspects of their proliferation into wider society were perceived as a risk to social order, even while an armed citizenry continued to be valued for its contribution to civic defence. As Robert C. Davis has pointed out, the development of this armed citizenry (to which should be added the development of the militias in particular) shifted long-standing social structures of power.[85] These shifts were, however, multidimensional. While the wide use of wheellocks and the framing of legislation around them certainly empowered both the highest elite (exempt from most gun laws) and those willing to operate outside the law, the rise of the handgun also empowered gunmakers, maintainers, transporters and dealers. Within these changing dynamics, gun laws and their enforcement came to be a site for complex negotiation between regimes and citizens. High-ranking individuals enjoyed significant privileges in terms of obtaining exemptions from the law; more generally, it came to be accepted that in certain circumstances carrying a gun was a legitimate means of self-defence. Nonetheless, as the anonymous treatise shows, there was scepticism about whether guns were in fact effective in responding to attack. Licensing systems offered one means of dealing with gun problems, but they depended on the presence of reliable staff who would issue licences in accordance with the law, and, moreover, could not prevent a weapon licensed for a legitimate purpose (such as the management of livestock) from being turned to illicit ends.

Having promoted the circulation of firearms for civic defence, the Italian states had to persuade citizens that they should respect the laws and use only permitted types of gun in permitted circumstances. Given the evident concerns about their ability to protect citizens from the banditry that plagued Italy in the second half of the sixteenth century, that was not easy. Indeed, one method employed to tackle violence in the 1570s arguably encouraged it. This was the practice of paying bounties: a Venetian provision of 1570 imposed strict bans on firearm use but offered a reward of 600 lire to those who delivered an offender to them, dead or alive.[86] (The monies would be recovered, where possible, from the offender's estate.) In a society riven by feud and vendetta, this was a deeply risky approach and might easily backfire. This was also the period in which restrictions on gun manufacturers were briefly attempted. In the last three decades of the sixteenth century, papal policy oscillated between stringent bans (including, in the 1570s, restrictions on production) and more liberal approaches.[87] As the contrast between Catterino Zen's approach and that of the anonymous author shows, while there was a strong sense among governing officials that there were problems with gun violence, there was no consensus on how to tackle it. Indeed, the two alternatives set forward here in the immediate aftermath of the firearm revolution—relax the law and allow gun ownership as a means of self-defence, or tighten regulation in the hope of removing the most risky weapons from society—have recurred in debates to the present day. Gun violence was, moreover, not only an Italian problem. Across Europe, high-profile shootings were on the rise.

PART III

Beyond Italy

CHAPTER SEVEN

Assassinations

AS WE SAW in previous chapters, guns that could be concealed were subject to particular regulation. Motivations for these bans were rarely spelt out, but there was a clear risk that small hidden guns might be used in assassination. The 1584 shooting of William the Silent, prince of Orange, is often claimed as a first, but here I show that firearms were seriously considered as a weapon in political murder for half a century beforehand. In this and the following two chapters, I set the Italian cases in a wider European landscape that helps to contextualize some of the decisions made on the peninsula and shows the broader impact of the firearm revolution. While there were important differences in culture across the continent, including in relation to guns, Europe was a connected space. Ruling families intermarried: Catherine de' Medici was queen of France; Margaret, duchess of Parma (an illegitimate daughter of Charles V) at one point served as governor of the Low Countries. Many military commanders moved between different European theatres of war, to the point that in the 1540s a Venetian diplomat could point out that there were no Spaniards on Charles V's secret war council, only Italians.[1]

This chapter assesses six cases of planned, attempted or successful assassination that involved firearms, stretching from the 1530s to the 1580s and from northern Italy via France to the Netherlands and England. In five of them, the gun was the offensive weapon; in the sixth the target's bodyguard had guns, and witness testimony sheds light on the limits of firearms in defence. Guns featured in a three successive assassination attempts in the years 1569–72, news about which may have both encouraged their use and prompted a revived hostility towards firearms. The accounts of these incidents together reveal a growing confidence in the use of handguns for shooting from (relatively) longer distances—raising questions about their efficacy in defence—but also a renewed concern and a reiteration of the idea of guns as diabolical, especially in the context of religious conflict. Unlike the criminal records, which typically record little qualitative detail of why gun users made certain choices, the assassinations sometimes prompted fuller comment. Together, they add further, and more international, context for the developments in gun control discussed in chapter 6. Assassinations also shed light on firearms in relation to an important question about early modernity: the rise of the individual. As we saw in previous chapters, handguns' efficacy on the battlefield depended on their deployment by groups of soldiers. In the early years of the sixteenth century, the lone gunman was limited by the range and accuracy of the available weapons. Over time, however, that changed.

Comparatively speaking, the framework for gun control in north and central Italy was remarkably similar to that in England. There, in the first half of the sixteenth century, gun control measures included the imposition of property qualifications for handgun ownership, licensing, and a particular focus on the control of short weapons, but also significant exemptions for country dwellers and those living near the coast or the border with Scotland, with an eye to personal protection and national security.[2] In Germany, there were likewise restrictions on the carrying of loaded weapons within the city walls; in the countryside there were restrictions on gun use in hunting. Nonetheless, an Imperial decree of 1530 permitted guns to be kept at home and looked positively on the practice of shooting matches. In the later sixteenth and seventeenth centuries, gun laws in the German countryside came to be focused increasingly on the problem of poaching.[3] This emphasis is less apparent in northern and central Italy, where the high rate of homicide and banditry made questions around personal protection from gun violence and the use of guns in self-defence more immediate than elsewhere. In the Netherlands and France, meanwhile, civil wars made any prospect of disarming the wider population highly implausible.

Furthermore, there were many political and cultural differences in attitudes towards assassination. Extrajudicial political murder was widely practised by rulers and their rivals in Italy (unlike, say, England where the Tudor regimes generally preferred the appearance of judicial process). One need only think of the infamous examples of the 1478 Pazzi Conspiracy in Florence or the various killings associated with the Borgia family during the papacy of Alexander VI (1492–1503). It is, therefore, not surprising that we find some notable early attempts at assassination by firearm on the Italian peninsula. Guns certainly afforded some advantage to an assassin in that they offered the possibility of killing at a distance. On the other hand, they were less reliable than other methods. Like bandits, at least some assassins chose to carry a pair of weapons; this may have been the motivation for the development of the illegal four-barrelled gun mentioned in chapter 1. Other assassins did not opt for guns at all. None of the three Medici murders of the 1530s and 1540s, for example, involved a firearm. Ippolito de' Medici was killed with poison, Alessandro de' Medici with a sword, and Lorenzino de' Medici with a dagger. A plot involving gunpowder did feature in the course of this vendetta, but the plan was to explode a chest of gunpowder: more akin to a bomb than a gun.[4] In some cases, the temptation to evoke the assassination of Julius Caesar may have been a factor in the choice of method: in others, poison allowed for deniability. The gun, however, offered new possibilities.

Pakington and Pole

The first European political assassination by handgun is sometimes said to be that of Robert Pakington, who was shot dead in London in November 1536. His murder coincided with the rebellion then shaking northern England known as the Pilgrimage of Grace. Pakington was a member of Parliament and a prominent religious reformer with links to Henry VIII's chief minister Thomas Cromwell, which made him a plausible target for Catholic rebels. Despite the promise of a generous reward, his killer was not found. Rumours attributed the murder to the bishop of London, in one case, and the dean of St Paul's Cathedral, in another. An alternative version had a known felon confessing. Whether or not the killing was religiously or politically motivated in fact, it was certainly accorded those motivations in subsequent Protestant accounts, including the influential *Acts and Monuments* of John Foxe, a Protestant martyrology that gained wide distribution: at one point every parish church in England was instructed to acquire a copy.[5]

While it is hard to be certain about the detail of Pakington's murder, the idea that handguns might be useful for the purpose of assassination was certainly circulating at this time, not least in relation to an unsuccessful plan by the English Crown to kill Cardinal Reginald Pole. The cardinal, who had been raised to that rank in 1536, was not only a religious opponent of King Henry VIII but had a claim to the throne via Plantagenet ancestors. He had left England for exile as Henry was preparing to break with Rome in 1532, and spent extended periods in northern Italy, studying at the University of Padua (located in Venetian territory). By 1537, when discussions about his assassination were taking place, handguns were well established at the English court. Italy had been a notable source of early technological transfer and imports. In 1512, Peter Corsi, an Italian, had been paid by the English Crown for 420 handguns.[6] The Bardi and Cavalcanti company, which had a close relationship to the Medici rulers of Florence and acted as an informal diplomatic channel to the English court, was responsible for the import of handguns and arquebuses to England in the same decade.[7] These transactions should be seen in the context of the long-standing relationship between the English court and those of northern Italy, which also saw exchanges of horses, hawks, hunting hounds, honours such as membership of the Order of the Garter, commissions for Italian artists, and a notable number of English students enrolling in the University of Padua.[8] After the break with Rome in 1533–34, England maintained diplomatic relations with Venice, a link that undoubtedly facilitated the substantial English order of Brescian guns documented in 1544, and Henry himself took shooting lessons.[9]

It was against this background that in 1537, Sir Peter Mewtas, a gentleman of Henry VIII's Privy Chamber, was apparently sent to assassinate Cardinal Pole. The details of this affair emerged only the following year, during interrogations concerning the so-called Exeter Conspiracy (in which relatives of Pole were accused of trying to overthrow the king). Morgan Wells, a servant, had apparently said openly that he 'would kill with a hand-gun Peter Meotes or any other whom he should know to kill the cardinal Pole, and that he was going over seas for that purpose'. The interrogation process, moreover, turned up a rumour that Mewtas planned to kill the cardinal 'with a hand-gun'.[10] If nothing else, this confirms that in the aftermath of Pakington's murder the idea of assassination by shooting was thought credible. It was, moreover, particularly credible in the specific case of Mewtas, who played a significant role in the rise of handguns in England. The same year he and a fellow member of the Privy Chamber were granted a licence to form an Artillery Company, with specific reference to 'longbows, cross-bows and handgonnes'; in 1543, Mewtas was given a licence by the Privy Council to levy one hundred arquebusiers.[11] Neither Mewtas's assassination mission nor further schemes in 1538 and 1546 were successful, however.[12] In avoiding attacks in the Veneto, Pole may have benefited from his local connections. These included, in Brescia, Giovanni Battista Porcellaga, whose arms brokering was discussed in chapter 3. In the Porcellaga account books for July 1553, Pole (who was then residing at the Benedictine Abbey in Maguzzano, near Lake Garda) appears as the recipient of several expensive birds, including a pair of guinea fowl and a pair of peacocks, one of them a rare white example.[13] This was a small world, and by that time Pole's prospects were looking up as the Catholic Mary Tudor saw off rivals to take the English throne.

Pier Luigi Farnese

Another individual from the Brescian records links us to our second case-study in assassinations. The investigation that followed the September 1547 murder of Pier Luigi Farnese, duke of Parma and Piacenza, provides important evidence for how guns were and were not used in violent incidents. The ducal household was very familiar with up-to-date weaponry. Pier Luigi had hired various military experts for his household, including, as master of arquebuses, Venturino da Chino (of the Brescian family), at a generous annual stipend of 5,298 lire; Venturino had facilitated a purchase of four thousand Brescian firearms for Pier Luigi the previous year.[14] The assassination took place shortly after lunch on 10 September.[15]

The duke was in his apartments in a citadel that should have been defensible. His bodyguard was armed with wheellock firearms, but the conspirators had guns too, and used them to force their way into the citadel. They did not, however, choose to shoot Pier Luigi, but to stab him. That may reflect the risks of shooting in an indoor space, but cultural factors may have been at work too. To kill by stabbing, with multiple assailants, echoed the assassination of Julius Caesar, as noted above. A decade previously, Lorenzino de' Medici had claimed the mantle of Brutus in defence of his murder of Duke Alessandro de' Medici, likewise carried out with an edged weapon.

The witnesses included multiple members of the Farnese household, including the bombardier, grooms, the butler and the steward. One groom, Nicola Fornari, was outside the citadel eating when he heard 'certain arquebus shots and cries of "arms, arms"'. Nicola explained that he and his companions went 'out of the lodgings . . . at the door of the chamber was Fabio [da Sassoferrato, another groom] . . . with an little wheellock still [in hand] and we tried to enter the citadel but could not enter'.[16] Fabio himself told the inquiry: 'I had an arquebus and then I left the arquebus and took a shield, thinking to go inside.'[17] It is notable that he saw these as alternatives rather than complementary weapons and, moreover, that he considered the defensive equipment to be the priority.

Giulio de Salvetis, the bombardier, was alerted to an incident by 'great noise and the firing of arquebuses'.[18] Like the other witnesses, he found himself locked out of the castle. He 'gave the arquebus to one of his servants . . . then he sent him home to get balls and powder because he wanted to fire a culverin (*colobrina*) that was on the piazza to break the door down; and he began to take arquebus shots at the windows of the citadel'.[19] Inside the citadel, the conspirators now pleaded with the crowd not to shoot them. As Giovanni Francesco Aloisi de Casal Sanvas (a household member of Sforza Pallavicino, who would later lead troops for Venice) explained:

> They let him [the murdered Pier Luigi] stay [hanging out of the window] and then [Francesco Maria Galasso] started to exhort with their cries the people who were on the piazza of the citadel with guns, not to shoot the arquebuses at them, because they had done it to liberate the fatherland from the hand of that tyrant.[20]

Turning back, Giovanni Francesco encountered his patron, Sforza Pallavicino, 'on a horse, *archabusino* in hand'.[21] There were lots of these small guns about. Giovanni Battista Giovanni de Lucianis, another witness, saw Agostino Lando, one of the conspirators 'on a small horse, with a small wheellock arquebus and a sallet [helmet] covered in velvet on his head, with about ten or twelve of his men around him, all with their little wheellocks in hand'.[22]

Giovanni Francesco Cesi[us], another of the murdered duke's servants, saw the conspirators from the window of his chamber, which looked out onto the piazza inside the citadel. They had 'large partisans [polearms] and arquebuses and diverse types of arms, with many men behind them'.[23] Cesi's testimony gives important detail about how the guns were used in practice: as the castle was sacked, the assailants stripped the servants at gunpoint, 'pointing the arquebuses at our chests', to see if they had money.[24] This was the typical robber's use of the firearm to threaten, but at other points during the attack guns were used to kill. Ludovico de Sexto de Frigoli, an official of the ducal household, witnessed the drawbridge being raised, 'and there was Count Agostino on a horse . . . with certain others, who with an *archibusetto* killed a guard'. (This is perhaps the same incident reported above by Giovanni Francesco Aloisi de Casal Sanvas.) Ludovico and a servant named Salvator headed upstairs (towards the ducal apartments), armed respectively with a *spetto* (a spit, or skewer) and nothing at all. Halfway up the stairs, they found two of the conspirators, Gianluigi Confalonieri and Giovanni Anguissola, who told Ludovico to 'save his life', while one of their men hit Ludovico over the shoulders with the barrel of his gun. Seeing the duke dead, Ludovico fled to the kitchen.[25]

Piecing together these accounts, a rounded picture of gun use in the course of the attack emerges. The conspirators and at least their immediate entourage were armed with small wheellock guns. In at least one case these were used to kill. In another, a gun was used as a non-lethal club. Firearms were also used to threaten in the course of robbery. The conspirators in turn faced attack from men like the bombardier Giulio de Salvetis, who began shooting at the citadel windows; they had to plead with the crowd not to shoot them, which makes clear that more people than just Giulio were outside with guns. If firearms were not the weapon of the principal murder here, they certainly played a role in facilitating the attack. They were not, however, so useful in defence.

Nor were guns a feasible option in all scenarios. Captain Francesco Bibboni, who assassinated Lorenzino de' Medici in Venice in 1548, used a dagger.[26] His own memoir of the event makes clear his familiarity with guns: he narrates how during a previous job he had seized arquebuses from a target and explains that he had fought with the Imperial army the previous year in Germany, military experience that would have acquainted him further with firearms.[27] As Bibboni fled north after the murder, he became suspicious of a man equipped with three wheellocks, one large and two small.[28] Such guns were clearly available. In terms of his weapon choice, the limited range of a small wheellock combined with the risk of misfiring may have been a consideration: if he had in any case to get close to his target then a dagger was arguably more reliable, especially in wet weather (he killed Lorenzino in February). In the circumstances, Bibboni was unlikely to get more than one chance, and on his own account bided his time and did not take risks, which might explain his avoidance of guns. Even the dagger he used was a banned weapon in the city, he wrote, and could get a user sent to the galleys.[29] Moreover, as we saw in chapter 6, the Venetian authorities were actively arresting individuals found with illegal firearms around this time. Even if Bibboni had judged that gun use would be a practical option, perhaps the gun laws and consequent risk of arrest worked as a deterrent.

The miracle of Carlo Borromeo

Twenty years later, the city of Milan was the setting for a further attempt at assassination by firearm. By this time, the Wars of Religion had begun in France and, amid wide talk about the use of assassination to political ends, already one prominent figure had been shot.[30] This was François, duke of Guise, who was killed in February 1563. We will return to François, whose wife was Anna d'Este of the ruling family of Ferrara. Suffice to say that the shooting of Carlo Borromeo, which took place in October 1569, perhaps had some relatively recent inspiration. At the time of the attack, Borromeo (1538–84) was the archbishop of Milan and a prominent figure in Catholic reform as the church responded to the challenge of Protestantism in the years after the conclusion of the Council of Trent. Borromeo had angered members of the Umiliati, a religious order, with a series of reform proposals the details of which need not concern us here. One friar, however, was sufficiently disgruntled to turn to violence, and his story was recounted in several texts written in the late sixteenth and early seventeenth centuries as Borromeo was considered for sainthood. The earliest is a 1591 account by the Mantuan author Giovan Battista Possevino.[31] Having outlined the conspiracy, which involved several individuals, he describes the process by which the shooter, Girolamo Donato, nicknamed Il Farina, robbed some church silver. Heading to Mantua, Il Farina sold part of his loot and pawned the rest, thereby obtaining a 'good sum of money', with which he went to the 'land of the Grisoni'—that is, north towards Switzerland—and 'equipped himself with small wheellock arquebuses' before heading to Milan 'to commit the wicked excess'. (A later account specifies that he went to Lugano, and that one of the wheellocks was somewhat larger than the other; either way, a second gun would have allowed for a second shot if the first misfired.[32]) Initially, Il Farina's plan was to rent a house on the route that Borromeo usually took to reach the church of San Barnaba, intending to 'shoot [Borromeo] with an arquebus, and then get on his horse' and make his escape 'along certain uninhabited little roads nearby'. Borromeo, however, changed his usual routine and did not visit San Barnaba. Il Farina therefore decided to go and kill him in the archbishop's palace itself. There, Borromeo was accustomed to praying every evening with his household in a little room: at this point his chapel was still under construction. On Il Farina's first attempt, he got to the door of the room 'to bring to fruition his diabolical plan', but then saw Borromeo in the company of Cardinal Crivello, and—concerned to avoid hitting Crivello—abandoned the attempt. Finally, however, on another evening and through a crack in the double doors, Il Farina took aim at Borromeo, shooting him in the back.

Appropriately for the occasion (and at this point some scepticism as to the accuracy of the account is required), as these events were playing out a choir was singing a motet with the words 'non turbetur cor vestrum, neque formidet'—that is, 'let not your heart be troubled, nor be afraid'. Borromeo continued with his prayers while Il Farina made a run for it, retreating to his brother's house where he holed up in the attic, claiming to his brother that he was hiding out for fear of being imprisoned, having quit his religious order. 'It was', wrote Possevino, 'a most manifest miracle, and evident sign of the providence that Our Lord holds towards his servants, that the Cardinal [Borromeo, by his future title] did not die: he, having finished the prayer, stayed immobile on his knees for more than a quarter of an hour, firmly believing (as he later confessed) that he had been shot through'. The gun, Possevino explained, had been loaded with one full-size ball and three small ones: the largest hit Borromeo in the back but left only a little red mark and some swelling; one of the smaller balls cut through his vestments to his undershirt, while the others hit a wall and a piece of wood. The main shot left a mark on Borromeo's body that was visible even after his death, witnessed personally by Possevino. The nature of the miracle is also a matter on which modern research might shed some light. We know from Possevino's account that Borromeo was wearing several layers: an undershirt, a doublet, a cassock and a rochet. Multiple layers of silk have long been used to protect against stabbing, and to some extent can stop bullets.[33] It is possible that liturgical vestments may have acted as a sort of body armour. In the absence of such scientific awareness, however, it is not surprising that Borromeo's survival was regarded as miraculous. Later versions of the story add further colour to the tale, including a comparison of Farina to Judas (though Farina was paid forty rather than thirty pieces of silver for his betrayal),[34] and the details that he secured the funds specifically from a Jewish moneylender, and that he did so with the licence of the rulers of Mantua.[35] One account specifically refers to the 'malign suggestions of the Devil' in the affair, unsurprising in context but reinforcing early ideas of the gun as a diabolical weapon.[36]

The Umiliati were disbanded in 1571, and Borromeo's survival was one of the miracles cited when he was first beatified (in 1602) then canonized (1610). With the beatification process came not just texts but visual depictions too (if images were produced earlier, I have found no evidence of them), and in this context, the gun entered the iconography of modern martyrdom. As we will see in

FIGURE 7.1 Giovanni Mauro della Rovere detto 'Il Fiammenghino', *The Blessed Charles survives the arquebus attack unharmed thanks to Divine Providence* (*Il beato Carlo sopravvive illeso all'attentato con l'archibugio grazie alla Divina Provvidenza*). Milan, Venerabile Fabbrica del Duomo, 1602. Photograph: Foto Milano, inv. FM B 357. Photographer unknown. © Civico Archivio Fotografico, Comune di Milano. Si resta a disposizione degli eventuali detentori di diritti che ad oggi non sia stato possibile rintracciare. We remain at the disposal of any rights holders who have not been traced to date.

chapter 8, guns rarely featured in paintings of biblical scenes, but here the impious weapon of the textual sources found its way into visual culture, aimed against the reformer and saint. This was one means by which the negative associations of firearms were reframed for new generations. The shooting scene appears in the 1602 series of *quadroni* (large paintings) produced for the Duomo in Milan, which are hung to mark the saint's day each November (Figure 7.1). A 1610 engraving likewise shows the scene (Figure 7.2); a fresco by Giovanni Mannozzi may be found at the church of Madonna dei Monti, in Rome (dating to the 1620s), and one by an unnamed artist in the church of San Michele and the Madonna of Biorca at

FIGURE 7.2 Alberto Ronco, 'Assassination Attempt on Charles Borromeo', from Cesare Bonino, *Some Remarkable Deeds of the Blessed Charles Borromeo* (*Nonnulla Praeclara Gesta Beati Caroli Borromaei*), Milan, 1610. Biblioteca Trivulziana, Milan. © NPL—DeA Picture Library / Bridgeman Images.

Grailè in the northern Alps (also dated to the seventeenth century) (Figure 7.3). In the 1630s, Guercino produced a lively sketch of the scene (now in the Royal Collection), which probably informed the fresco painted by his pupil Antonio Bonfanti for the church of San Carlo in Ferrara.[37] Here, moreover, the images of guns

FIGURE 7.3 Unknown artist, *St Charles Borromeo, having survived an attack* (*San Carlo Borromeo sopravissuto ad un attentato*), 17th century. Chiesa di S. Michele e della Madonna di Biorca. Photo: Rita Guglielmi via Alamy.

underline the modernity of the saint: in a context where Protestant churches no longer created saints this was a distinctively new Catholic image.

The assassination of the earl of Moray

If the Archbishop Borromeo was miraculously saved, however, only a few months later came the first assassination of a European ruler with a gun. This was the killing of James Stewart, earl of Moray and regent of Scotland, on 23 January 1570 in the small Scottish city of Linlithgow. Moray, an illegitimate son of King James V of Scotland, had become regent for the young king James VI following the abdication of Mary, queen of Scots, in 1567. The shooter in this case was James Hamilton of Bothwellhaugh, a supporter of Mary whose family were long-standing enemies of Moray; he targeted Moray as he was passing in a procession. Contemporary accounts make it clear that this was a plot by the Hamiltons collectively, not an individual scheme. Robert Birrel explains in his diary how

> the Earl of Moray, the Good Regent of Scotland, was slain in Linlithgow by James Hamilton of Bothwellhaugh, who shot the said Regent with a gun out at a window, and presently thereafter fled out at the back, and leapt on a very good horse, which the Hamiltons had ready waiting for him; and, being followed speedily, after that spur and wand had failed him, he drew forth his dagger, and struck his horse behind; which caused the horse to leap a very broad stank; by which means he escaped.[38]

An anonymous diarist made a note on 'the slaughter of James erle of Murray regent', confirming the identity of the assassin, and noting that he 'lay at await to commit this slauchter, in the bischope of Sanctandrois ludgeing in Linlithgow'. On this account, Hamilton's escape was facilitated by 'diverse personis his assistaris'.[39] The contemporary memoirs of James Melville, a Scottish diplomat, suggest that the shooting could have been avoided if the earl's advisors had acted on intelligence received.[40] To these sources may be added a somewhat later historical account, that of David Calderwood, written in the seventeenth century, who described the assassination 'with a hacquebutt, through a tirleis [shuttered] window, from a stair wherupon were hung sheetes to drie, but in truthe, to hide the smooke, and mak the place the lesse suspected'.[41]

It is not clear from Calderwood's account whether the smoke he refers to was the smoke following the shot (which would have been relevant in identifying the location of the shooter after the event), or whether he thought Hamilton was using a matchlock, which would have required a smoking match-cord in advance. Given the ubiquity of wheellocks by this time, the former seems more likely: none of the sources comment on the specifics of the gun. In terms of the assassination's success, the weapon was only one part of the picture. As significant were the powerful backers of the plot and the carelessness with which Moray's household had treated warnings of an attack that were detailed enough to include the location from which it would be carried out. In any case, the method as established avoided the need for the assassin to get up close (as was necessary when a very short wheellock was used); it also improved the chance of escape given that the gunman might already be at some distance from any guards.

Moray is now often claimed as the first head of government to be assassinated with a handgun. The novelty does not seem to have struck contemporaries, who reported the gun use in rather practical terms. Perhaps after so many attempts it seemed only inevitable that eventually a gun would be used in such an attack. Moreover, this was not the first time that a person close to the Scottish royal family had been shot and killed. François, duke of Guise and uncle of Mary, queen of Scots, had suffered just such an attack seven years previously, in February 1563, in the context of the Wars of Religion in France.[42] Guise had at this time been the commander of the French Catholic forces, and his murder by Poltrot de Mérey had the effect of interrupting hostilities.[43] Plots against Guise had evidently been in train for some time but had been given new impetus by the 1562 massacre of Protestant worshippers at Vassy by Guise and his men.[44]

Shootings in the French Wars of Religion

By the time the wars in France began, guns were extensively used in the military, making up a high proportion of the Huguenot infantry companies, which had a balance of 80:20 arquebusiers to pikemen, alongside some arquebusier-only units.[45] Gun violence was also prominent off the battlefield, in events such as the murder of Marguerite de Hurtelon, a Huguenot noblewoman who was shot 'five times in her breasts with a pistol', and whose case featured in a list of examples of Catholic violence compiled by Protestants.[46] This was not to say that cultural attitudes had changed: the surgeon Ambroise Paré, whose treatment of gunshot wounds I discussed in chapter 1, was still working in the aftermath of the Battle of Dreux in 1561. 'I treated fourteen of them in a single room, all wounded by pistol shots and by other diabolical firearms,' he wrote, reiterating the persistent view that these weapons were the Devil's work.[47] Immediately prior to Guise's own shooting, his men had fired on Calvinist worshippers who were meeting in a barn at what became known as the Massacre of Vassy.[48] He himself was shot while returning home by a member of his own household. This man had loaded his 'pistolle' with three bullets and shot from 'six or seven paces'.[49] Thomas Smith (the English ambassador) wrote that the 'dag' (the term used in English for such small pistols) was made 'so strong that it received three pellets and three charges in one chamber'.[50] This was a similar method to that employed later in the decade by Il Farina to target Archbishop Borromeo, and reflects the limitations of the very short concealable pocket guns: the assassin had to be nearby.

By the 1570s, however, the range of longer handguns was sufficient to give potential assassins the confidence to use more sniper-like approaches. James Hamilton's attack on the earl of Moray was closely followed in France by the first, unsuccessful attempt to assassinate Gaspard de Coligny, seigneur de Châtillon and admiral of France, who was involved in commissioning the Guise shooting).[51] Coligny's injury, widely perceived to have sparked the subsequent St Bartholomew's Day Massacre, was caused by a shot from a window that hit his hand and arm. The details of the events have been pieced together by a number of historians from sources that are not always consistent in their details. However, some fundamentals are clear: the assassin shot from a window and was concealed behind either a washing line or some other type of drapery; he had planned carefully, arranging more than one getaway horse. An unexpected movement by Coligny meant the shot missed his vital organs, but he was murdered in any case two days later (Figure 7.4).[52]

Historical discussion of this event has most commonly focused on the question of who was ultimately responsible, with possibilities ranging from low-ranking members of the Guise network acting autonomously to King Charles IX and his mother Catherine de' Medici. Whatever the solution to the mystery, a good case can be made for familiarity with events in Linlithgow on the part both of the assassin and the writers who subsequently described events. These include the shooting from a window, the fact that the target was accompanied (in Coligny's case by an entourage, in Moray's case by a procession), the shooter's concealment (behind a curtain, sheets or shutters). Like Il Farina, this shooter apparently loaded three shots.[53] Like Borromeo, Coligny was not so badly injured that he could not speak, and reports have him pointing up to the window and saying, 'See how good people are treated in France!'[54] As in the previous cases, the shooter was initially able to flee the scene,

FIGURE 7.4 Franz Hogenberg, *Death of Admiral Gaspard II de Coligny* c. 1572. Bibliothèque Nationale de France. © Archives Charmet / Bridgeman Images.

in this case evading capture altogether. It is assumed that he was Charles de Louvier, sieur de Maurevert.

News about these events circulated across Europe thanks to pamphlets such as the anonymous *De furoribus gallicis* (On the French Outrages; in fact compiled by François Hotman), which was published in four languages and enjoyed at least eight editions.[55] Taken together, these three incidents in three years may well have contributed to the burst of hostility to firearms apparent in the European law and literature of the 1570s. The assassinations provide important context for the attempts in the Papal States to ban wheellock guns entirely (not just use but production too) during this decade.[56] They also provide context for the English literary debate—many years after the use of guns had become normalised within European armies—about their proper role in wartime. This involved works such as Peter Withorne, *Certaine Wayes for the Ordering of Souldiours in Battelray, Plottes for Fortification, How to Make Gunpouder, &c.* (1573) as well as Barnabe Rich, *A Right Exelent and Pleasaunt Dialogue, betwene Mercury and an English Soldier* (1574), and the first translation, by Withorne, of Machiavelli's *The Arte of Warre* (1573). In the face of social change, argument persisted into the seventeenth century.[57] Meanwhile, rulers took precautions. Elite bodyguards were routinely armed with arquebuses, including that of Henry of Valois when in 1574 he made his entrance into Krakow for his coronation as king of Poland, and that of the duke of Moscow.[58]

The assassination of William the Silent

None of the debate, however, put an end to the usefulness of handguns for murder, and 1584 saw the assassination of William the Silent, prince of Orange. A leader of the Dutch Revolt who had been outlawed by Philip II of Spain, William had a price of 25,000 ducats on his head. He was shot with a pistol by Balthasar Gérard, a twenty-five-year-old Frenchman, who had infiltrated William's household by offering the Orange faction access to secret Spanish intelligence.[59] His ploy worked: the bullet holes may still be seen at the Prinsenhof in Delft. This murder is sometimes regarded as a first, and in technical terms it may be the first assassination by handgun of a head of state. As this chapter has shown, however, handguns had been deployed in high-level political murder certainly for two decades and possibly for five. William's killing is better understood as the culmination of that process.

Gérard was not the first assassin to target William, nor was he the first with a gun. An earlier attempt had been made by Juan de Jáuregui, who shot William in the head on 18 March 1582, having pretended to be a petitioner to gain access to him. As in the case of Archbishop Borromeo, close range was important to a successful shooting with a small firearm. Here, however, the assassin was let down by his technology: the pistol exploded in his hand. The shot hit William in the neck, but he survived the injury. Jáuregui was killed on the spot, and two associates were subsequently executed.[60] The man who had commissioned the attack, Jáuregui's employer Gaspar de Añastro (who appears to have been motivated primarily by a promised reward), escaped punishment.

William lived another two years before he was finally assassinated by Gérard. Once again, this was a shooting at close range, accomplished on the stairs of the royal residence, and like the murder of the duke of Guise required infiltration of the household in order to gain access to the prince. Gérard obtained his pistol from another member of William's entourage, saying that he needed it for self-defence while travelling, arguably the most socially acceptable reason to carry a firearm, but one that ensured guns were available for less benign purposes.[61]

Conclusion

The assassinations (and attempts at assassination) of the sixteenth century provide a distinct context in which to consider attitudes towards firearms. In the contemporary accounts of these incidents, the guns are noted but not particularly highlighted. Thanks to the firearm revolution, handguns had become normal, routine objects. If these incidents are picked out as firsts, then that is with hindsight concerning more modern gun violence. That is not, however, to say that they had no impact at the time. The more stringent papal policy towards guns of the 1570s (including restrictions on production) or the stop-and-search policy introduced by the Este rulers of Ferrara in 1573 might plausibly have been influenced by the wider climate. Members of the papal court were no doubt concerned at the attack on Borromeo, while through their marriage alliances the Este were active players in the French Wars of Religion. Temporary tightening of policy did not, however, prevent further attacks and indeed the eventual loosening of gun control laws as governments opted to allow the use of firearms for self-defence and to arm their own militias. Moreover, without wishing to overstate the similarities between these cases, a well-prepared assassin would no doubt wish to learn from prior experience. On the other hand, so did those at risk, and after William the Silent it was a full two centuries before another head of state suffered the same fate.

Along with spiralling factional violence and debates about the stringency of papal policy, these assassinations provide further context for the publication of the anonymous treatise lamenting gun-toting cowherds. In addressing the problem of gun proliferation, the author emphasized the importance of multilateralism. Were one state alone to ban wheel-locks, people living in the border areas would feel vulnerable to attack from bandits who could easily obtain guns from the neighbouring polity. As we have seen, this was the case in the attempted assassination of Carlo Borromeo, where the firearms were purchased outside Milan. Some cross-border legislation was, in fact, established, albeit on a modest scale. In 1580 the duke of Ferrara agreed with the government of Bologna that bandits might be pursued three miles into Ferrarese territory, and that he would neither harbour them nor grant them safe-conduct.[62] The anonymous author, moreover, proposed that the pope, who (at least in theory) had a moral authority other princes lacked, should take the lead in securing a collective end to the production and use of wheellocks throughout the Italian peninsula, the target being not so much the users of such guns (whose motives he regarded with some sympathy, even while he was sceptical about the efficacy of a gun for self-defence against a surprise attack) as their producers and repairers, whom he argued should face the death penalty. The Italian states, he wrote, should ban foreign imports of wheellocks and collectively abandon the use of the weapon in military contexts in favour of the 'more secure', or 'safer', matchlock (*arma più sicura*). While this suggestion is consonant with much older ideas of the papacy as a peacemaker among the Italian princes, it is striking that the idea of a multilateral arms reduction treaty addressing the spread of the continent's most dangerous weapons could be conceived of in the sixteenth century, almost four centuries before any such treaties were in fact enacted.[63]

The assassination cases confirm what an individual might achieve, but also the importance of teamwork. The need for access to a suitable location and the potential for escape both depend on there being more than one person involved. Yet, on the other hand, the risk of the 'wretch with the gun', as Monluc had put it, killing a prominent political or religious leader was now all too evident. For a gun to be effective no longer required its use in a large military unit. Nor (as in the case of interpersonal violence) was its use primarily to intimidate: the act of killing itself could be used to intimidate others. The gun had come into its own as an increasingly reliable individual weapon, and one with potential to escalate conflict on a much larger scale.

CHAPTER EIGHT

Visual Culture

GIVEN THE PROLIFERATION of firearms into European society by the third quarter of the sixteenth century, one might expect to see a similar proliferation in the continent's visual culture. In fact, however, continuing ambivalence about the new technology was reflected in its representation. Sometimes guns retained their diabolical associations; on other occasions they were proudly on display. The fact that guns were common does not mean they were universally accepted. In Italian art, the search for the gun is largely a search for exceptions: the 'visual silence' Hale observed in relation to the Italian soldier in art prevails for the new weapon too.[1] Queries on the Cini Archive's database of the woodcuts used to illustrate popular Italian print turn up few depictions of firearms: when they do feature, it is in the context of contemporary conflicts.[2] In 1551/52, the Venetian artist Lorenzo Lotto painted a portrait of a Maestro Battista with his crossbow, now in the Capitoline Museums but originally produced in Ancona, on the Adriatic coast.[3] This was around the time of the papal siege of Mirandola, during which (as we have seen) significant numbers of firearms were in circulation; if Battista used a crossbow regularly it is more likely to have been for hunting than for warfare. Yet, to my knowledge, there is no equivalent portrait of a sitter from the middle ranks of society with a handgun. Outside images of contemporary warfare, guns are rare in both portraiture and religious art.

That was not the case across Europe, however. In both the Low Countries and the German states guns were more readily shown, even if they sometimes retained negative associations, reflecting a broader cultural contrast in the representation of warfare.[4] This chapter explores trends in the depiction of guns across western Europe, asking what we might learn about firearms from their portrayal in visual culture, and exploring how those portrayals fit with the evidence of the written record. It begins with a notable case from the fifteenth century, before turning to examples from north of the Alps, setting in context Titian's mid-century equestrian portrait of Charles V, which shows the Emperor with a wheellock gun. It then assesses the presence (or rather lack) of guns in paintings of biblical scenes, and in depictions of contemporary conflict, including with the Ottomans, before turning briefly to the art of the French Wars of Religion and Dutch Revolt. Towards the end of the century, artists working in England began to show guns in the context of colonial and imperial projects, a notable turning-point in the iconography, and one that pointed to a new, more positive narrative of guns.

Alexander the Great

We have already encountered a number of fifteenth-century portrayals of guns. These include the Florentine *cassone* panel depicting the Battle of Anghiari, and the French Book of Hours image of demons shooting at the risen Christ, along with the Pastrana tapestries depicting the Siege of Asilah. There are also some early examples of firearms in German art, as we will see below. To these should be added a remarkable 1460s tapestry, now in the Villa del Principe, Genoa (once residence of the celebrated sixteenth-century admiral Andrea Doria), and is a rare example in which firearms are incorporated into a classicizing narrative. The tapestry is one of two depicting the *Stories of Alexander the Great*, and shows several scenes from Alexander's life, including the conquest of a city. Produced in Tournai in the 1460s, it may have been designed by Jacques Daret, and was presented by Pasquier Grenier, a merchant, to Philip the Good, duke of Burgundy (Grenier is also thought to have supplied the Pastrana tapestries depicting the Siege of Asilah discussed in chapter 1).[5] Unlike those representations, however, these tapestries portray an ancient battle: although its figures appear in

FIGURE 8.1 ?Jacques Daret, *Histories of Alexander the Great*, c. 1460, Tournai. Tapestry: gold, silver, silk and wool. Approx. 40 m^2. Villa del Principe, Genoa. Photo: René Mattes / hemis.fr via Alamy.

fifteenth-century dress, one would not expect to find a firearm depicted. Still, among the numerous figures bearing fifteenth-century weapons such as bows and cannon is a single individual with a portable firearm. That is unusual, though not unique: a similar representation is to be found in a French miniature of c. 1480.[6] However, even more notable in the Tournai tapestry is the portrayal of the handgunner as Black (Figures 8.1 and 8.2).

It is a challenge to decode this representation. The inclusion may be an allusion to Alexander's conquest of the peoples of the world, although that does not explain the gun. Designs for further tapestries in the sequence (only two survive) suggest that it would have included at least two other Black figures with handguns; similar imagery appears in a contemporary sequence of the life of King Clovis in Rheims.[7] It seems unlikely that these were references to real people: the images are stereotypical, featuring the headband and earring characteristic of the Saracen's head, a common heraldic device; nor does the overall composition of the tapestry include other identifiable individuals. Jean Devisse and Michel

FIGURE 8.2 ?Jacques Daret, *Histories of Alexander the Great* (detail), c. 1460, Tournai. Tapestry: gold, silver, silk and wool. Approx. 40 m^2. Villa del Principe, Genoa. Photo: René Mattes / hemis.fr via Alamy.

Mollat suggested that the 'most likely' explanation for the presence of Black figures in these sequences was 'the firsthand observation of the presence of blacks in armies and a dependence on the African to heighten the marvelous and the exotic'.[8] A useful contrast here may be made with the frescoes showing the procession of the Magi in the Gozzoli chapel of the Palazzo Medici in Florence (now the Palazzo Medici-Riccardi). Produced a decade earlier, these include the image of a Black archer, whose facial features are sufficiently detailed and individual as to suggest this was a portrait, an interpretation supported by the inclusion of multiple portraits of the Medici family and household in the sequence. While there were no doubt Black people in fifteenth-century Europe who were familiar with firearms (a visitor to the Lisbon foundry observed Black workers there as early as 1494), the generic appearance of the tapestry figure points to a different interpretation.[9]

Devisse and Mollat say of the handguns in the Alexander sequence only that they point to 'a bit of sharp observation in a context of legend'.[10] Might we go further? Representations of Blackness in fifteenth-century Europe were varied. Numerous portrayals of the Magi (though not the Gozzoli chapel one) show one of the three Wise Men, or a page in his train, or both, as Black. In northern European art, St Maurice, a soldier martyred in Roman North Africa, also frequently appears as a Black man.[11] These positive Christian representations, which served a purpose in underlining the global reach of that religion, were, however, only one side of the story. There were also negative portrayals of Black figures in Christian art, notably some mid-fifteenth-century images associating Black figures with the Antichrist.[12] Given the wider negative literary and artistic culture around handguns, I am inclined to read the image in the tapestry as echoing the negatives. As we saw in chapter 1, a late fifteenth-century French Book of Hours showed demons wielding firearms against Christ. It is worth pausing here to observe that the demons are shown with dark skin, a conventional association of blackness with the Devil in medieval visual culture.[13] An association of diabolical firearms with unchristian Blackness seems the most straightforward explanation for the tapestry designer's choices.

The Low Countries, Germany, and Switzerland

Within a few decades, however, the narrative of European art would start to shift, especially in the Low Countries, Switzerland and the German lands, which were relatively more open than Italy to the portrayal of military subjects. A shooter loading his gun appears, for example, besides an image of a woman tending to an injured soldier in the miniatures of Bendicht Tschachtlan's *Bern Chronicle* (c. 1483).[14] While portrait representations remained few and far between, they began to be made. The earliest painting of an identifiable group of Europeans with handguns is the 1529 painting of a group of seventeen Amsterdam guardsmen, now in the Rijksmuseum, by Dirck Jacobsz.[15] Many of the city's militia companies commissioned paintings, of which the most famous is undoubtedly Rembrandt's *Company of Captain Frans Banning Cocq and Lieutenant Willem van Ruytenburch* (1642), better known as *The Night Watch*.[16] By the second quarter of the sixteenth century, gun use was increasingly normalized (though also highly regulated) in European cities, and participation in this martial culture essential to demonstrating citizenship and masculinity; in the Dutch context specifically, the late medieval militia guilds had been formalized into new civic militia.[17] In the early Jacobsz painting, however, the guns are shown relatively discreetly. This is not a group of men parading with this weaponry: the firearms are certainly not at the shoulder ready to fire, and the stand-out feature of the composition lies in its almost three-dimensional hand gestures.[18] Subtlety remains the order of the day in a 1531 painting attributed to Jacobsz's relative Cornelis Anthonisz.[19] In some of the seventeenth-century Dutch imagery, including one example to which we will come, firearms are more prominent. Outside the Low Countries, however, there is no equivalent of this tradition. Why did the Italian militias not have their portraits done? The stronger classical emphasis in art offers one explanation; the increasing aristocratization of the Italian ruling class through its 'flight to the land' (away from mercantile business in the cities) may also have been a factor.[20] Netherlandish art reflected a stronger bourgeois culture—also notable in portraits of merchants at work—that did not typically find a counterpart in Italy.

The most notable Italian exception so far as gunpowder weapons in general are concerned is Titian's portrait of Alfonso d'Este (r. 1505–34) with a cannon, later adapted by Battista Dossi and others (Figure 8.3).[21] Alfonso had arguably led the way in redefining aristocratic culture in northern Italy when it came to gunpowder weapons, and, as we saw in chapter 4, handguns featured in his court records as early as the 1510s and 1520s.[22] The duke's interest

FIGURE 8.3 Unknown artist, after Titian (Tiziano Vecellio), *Alfonso d'Este, Duke of Ferrara*, late 16th or early 17th century. Oil on canvas, 127 × 98.4 cm. Metropolitan Museum of Art, Munsey Fund, 1927, 27.56. Public domain.

in military technology was well known: he was happy to get his hands dirty overseeing artillery casting in his foundry, and enjoyed significant military success thanks to his efforts, notably at the 1509 Battle of Polesella, when his troops sank a Venetian fleet moored on the River Po. Dossi's version of the portrait shows this event in the background.[23] This portrayal was in distinct contrast to more classicizing imagery of rulers of the period, such as Lorenzo Costa's *Triumph of Federico II Gonzaga*, painted in 1522 for the Palazzo San Sebastiano in Mantua, now in the Sternbersky Palace, Prague: while some of the staff weapons shown by Costa are plausibly contemporary to the date of the painting, nothing is included that would compromise the overall imagery of Federico as a Roman commander. If this reflects broader trends in Renaissance art, it also avoids the difficult decision about how far to associate a ruler with the new and contentious military technology. As we have seen, it was not until the 1550s that guns began to be given in Mantuan ceremonial contexts.

Artists in the German lands, on the other hand, were rather more open to linking imagery of their rulers with the new technology. Between 1512 and the death of the Holy Roman Emperor Maximilian in 1519, a series of artists, led by Hans Burgkmair, produced designs for a large frieze depicting a triumphal procession. The Emperor hoped that the images would 'grace the walls of council chambers and great halls of the empire, proclaiming for posterity the noble aims of their erstwhile ruler'.[24] The panels show many soldiers and a large collection of gunpowder weaponry (Figure 1.9). More detailed imagery of firearms was already to be seen elsewhere, for example in Jörg Breu the Elder's *Roundel Depicting the Bohemian Battle* (c. 1515, showing the 1504 Battle of Wenzenbach), which includes a foreground image of reloading.[25]

In 1547, the Holy Roman Emperor Charles V enjoyed victory against his Protestant opponents at the Battle of Mühlberg. As we have seen, some luxury guns produced around this time incorporated Christian imagery in a turn away from earlier narratives of guns as diabolical. It was in this context that Titian painted the Emperor's equestrian portrait, showing Charles 'in the manner in which he went against the rebels'—that is, in the uniform of the Spanish light cavalry and equipped with a wheellock (Figure 8.4).[26] A similar gun, made by Peter Peck, is now in the Musée de l'Armée, Paris.[27] The date of this portrait falls within a decade of the 1553 change to the podestà's gift from crossbow to handgun in Mantua (see chapter 4). This was the point by which firearms had evidently been assimilated into court culture. By allowing himself to be

FIGURE 8.4 Titian (Tiziano Vecellio), *Emperor Charles V at Mühlberg*, Venice, 1548. Oil on canvas, 335 × 283 cm, Museo del Prado, P000410. Photo: ICP via Alamy.

painted with a gun, Charles associated it with the image of victorious Christian martial masculinity in the equestrian portrait. The painting, moreover, provided a visual counterpoint to the idea of firearms as the Devil's work. Just as Titian's depiction of Alfonso d'Este beside his cannon had emphasized the importance of that earlier military technology to the duke's power, now the delicate but lethal wheellock was incorporated into imperial imagery. By 1627, however, when Van Dyck (perhaps, the attribution is contested) came to produce the loose copy of Titian's Charles V that is now in the Uffizi, among the details to be dropped from the composition was Charles's gun. Indeed, Van Dyck transformed the ready-for-battle Charles into a relatively more civilian king: his armour is more suited to jousting, incorporating a rondel to support a lance.[28] Even as guns became the dominant technology in warfare, they were hidden from view.

Guns have also been somewhat hidden from view in the scholarship on the Titian portrait, which has tended to emphasize its metaphorical nature: it is not a battle painting, and scholars take different views on whether Charles's spear is intended to represent a sixteenth-century weapon. There is enough to allude to the Battle of Mühlberg, but not so much that this cannot also be a more timeless depiction of Charles as an ideal Christian soldier.[29] Only in a 2008 exhibition catalogue was the presence of the gun acknowledged.[30] Charles's portrait is, however, unique in the sixteenth century as a straightforward depiction of a European ruler with a handgun. True, when Arcimboldo produced his allegory of Fire in 1566, it alluded to the Emperor Maximilian II, with an image made up of objects including a cannon, a handgun barrel and a powder-flask, its forehead consisting of match-cord (Figure 8.5). Yet even there the gunpowder weapons were hiding in plain sight.

It is not until the seventeenth century and the court hunting portraits of Velázquez that we begin to see further depictions of rulers with their guns, such as that of Philip IV in hunting dress, painted around 1632–34.[31] These followed a wider incorporation of images of hunting with guns into court culture, such as the engravings of the hunt by Jan van der Straet (Giovanni Stradano), discussed in chapter 5.[32] A gun also features in a fresco of around 1606–10 by an unidentified artist originally in the Palazzo Caffarelli, Rome, and now in the Capitoline Museums. The hunter is posing on one knee, gun at the ready, accompanied by his dog.[33] The sequence as a whole celebrated the Caffarelli family's relationship with the Habsburgs, and borrowed from work by two artists with Low Countries connections, Paul Bril and Antonio Tempesta, a factor that may have favoured the inclusion of firearms.[34]

FIGURE 8.5 Giuseppe Arcimboldo, *Fire*, 1566. Oil on wood, 66.5×51 cm. KHM, Gemäldegalerie, 1585. © KHM-Museumsverband.

Wider imagery of the Battle of Mühlberg—in multiple media—emphasized the use of firearms by the army. Guns appeared on a portrait medal of Charles and Ferdinand with an obverse commemorating the capture of John Frederick I, elector of Saxony, now in the Frick.[35] A drawing by Giovanni Battista Scultori of Charles's

FIGURE 8.6 Unknown maker, majolica dish showing Battle of Mühlberg, Urbino, c. 1555–60. Tin-glazed earthenware, 43.5 cm diameter. Victoria and Albert Museum, 8926–1863. © Victoria and Albert Museum.

troops crossing the Elbe prior to the Battle of Mühlberg commissioned for Charles's ally Cosimo de' Medici in 1551 was the basis for an engraving by Enea Vico and in turn a majolica portrayal (now in the Wallace Collection), produced in Castel Durante, probably in the workshop of Ludovico and Angelo Picchi, 1559.[36] With its reloading arquebusier in the foreground, the image echoes earlier representations of shooters, including in the tapestry representations of the Battles of Pavia and Tunis, and inspired further copies (Figure 8.6). Collectively, these objects illustrate the widening presence of gun imagery across multiple media that might be displayed in a variety of domestic settings, including the studies and *Wunderkammer* that housed collections of valued objects to be shown to selected visitors.[37] In some cases, arms and armour might be a study's main feature. In his 1581 history of Venice, Francesco Sansovino noted the presence of several *studi di arme* in the city's noble houses, alongside studies dedicated to antiquities or music.[38]

Sansovino does not discuss their specific contents, but these would surely be plausible locations for the display of luxury firearms.

Firearms in religious painting

Despite the regular updating of visual representations of Bible stories to modern dress, early modern religious paintings very rarely showed firearms in this context. The Book of Hours of Charles d'Angoulême with its demon shooting at the risen Christ does not find an equivalent in paintings, so far as I have been able to tell. The closest example is a 1475 St Sebastian by Antonio and Piero del Pollaiuolo (in the National Gallery), which shows the saint being shot with a crossbow (and two further crossbowmen preparing to shoot).[39] Hans Holbein the Elder also showed crossbowmen in his 1516 St Sebastian.[40] The 1499 Göttingen altarpiece by Hans Raphon, now in the Sternbersky Palace, features a soldier with a crossbow in the scene of Christ's resurrection (notably the same scene in which guns were included in the Angoulême Hours).[41] Even crossbows, however, are rare in the iconography, and firearms rarer still. John Hale identified a handful: Joerg Ratgeb's 1517–18 *Resurrection* shows an arquebusier along with a crossbowman at the feet of the risen Christ; Joachim Patinir's *Baptism of Christ* (1510–20) includes an arquebusier listening to John the Baptist; Jacob Cornelisz van Oostsanen's *David and Abigail* (1520) shows soldiers in contemporary dress, including handgunners.[42] None of these are, of course, Italian. As St Barbara became the patron saint of gunpowder technologies, she was occasionally shown with a cannon, for example in the sixteenth-century altarpiece now in Palazzo Bellomo, Syracuse.[43] Votive paintings from the 1520s and '30s in the church of San Nicola at Tolentino (in the Marche region of Italy), presented to give thanks to God, made reference to contemporary shootings in much the same way as the images of the attack on Carlo Borromeo did.[44] In general, however, small arms found a place in religious art only in very limited circumstances. This can be explained not only by the perception of guns as diabolical. By the time that firearms had become sufficiently established as a military technology that artists might have thought to include them in the backdrop of a modern-dress religious painting, new considerations had arisen. The Renaissance fashion for humanizing Christian subjects had fallen out of favour: Counter-Reformation religious art was notably less humanist and more heavenly, emphasizing the saintly qualities of its figures. Many Protestants, on the other hand, disdained religious imagery to the point of iconoclasm. The wider ambivalence of literary and artistic culture is also apparent in contemporary writing on the Christian soldier, such as Antonio Possevino's 1569 *The Christian Soldier*, published in Rome, which discussed how the soldier should arm himself, but avoided any explicit reference to guns.[45]

Post-Reformation, the few exceptions to the no-guns rule appear in art from the Low Countries and especially in the work of Pieter Bruegel the Elder.[46] Bruegel's readiness to show firearms and other weapons in his paintings may reflect the more accepting local culture. His *Adoration of the Magi* (1564) includes a group of soldiers and a prominent and highly unusual image of a crossbow, just to the left of the Virgin and Child.[47] A crossbow is also shown in a *Resurrection* after Bruegel (dated after 1562), while his *Conversion of St Paul* (1567) features a crossbow and priming-flask, both prominent in the composition, while in the background a soldier is reloading a handgun (Figure 8.7).[48] It is a subtle inclusion, characteristic of Bruegel's attention to contemporary detail. One scholar analysing the painting discusses the pointed finger of the figure in the right foreground of the image with the green-plumed hat, describing its meaning as 'obscure'.[49] In fact, however, the tip of the finger precisely coincides with the background handgunner, which given the conventional use of the pointing finger or manicule in contemporary manuscript culture seems unlikely to be coincidence. Perhaps it alludes to Paul's future martyrdom. A gun is also included in the *Massacre of the Innocents*, painted in the last quarter of the sixteenth century by Pieter Brueghel the Younger and probably copied from a work of his father. Again, the weapon is not prominent. It is significant that these minor exceptions occur in Netherlandish painting and, moreover, in the work of a painter who during the 1550s and 1560s was not working directly on commissions for churches or religious institutions, rather for private residences, and whose own religious position was, to quote the editors of a recent volume on the subject, 'not easy to determine'.[50]

FIGURE 8.7 Pieter Bruegel the Elder, *The Conversion of Paul* (detail), Netherlands, 1567. Oil on wood, 108.3 × 156.3 cm. KHM, Gemäldegalerie, 3690. © KHM-Museumsverband.

While contemporary religious conflict was beginning to offer new opportunities for the depiction of firearms, as in the case of Carlo Borromeo described in chapter 7, in general, guns and biblical subjects did not mix.

Historical depictions

In contrast, guns continued to be shown in depictions of recent wars. As we have seen, they feature in both the Pastrana tapestries showing the fifteenth-century Siege of Asilah and the Avalos series of the Battle of Pavia. A series of tapestries showing the Conquest of Tunis were commissioned by Mary of Hungary to commemorate the 1535 victory of her brother, the Emperor Charles V. Like the earlier set showing the Imperial victory at Pavia, these were designed in the Low Countries; they were based on sketches by Jan Cornelisz Vermeyen, who had been commissioned to travel with the troops on campaign. A contract for the cartoons was issued in 1546, but the set of twelve (of which ten survive in Madrid) was not completed until 1554, when they were presented on the occasion of Philip II of Spain's second wedding to Mary, queen of England.[51] The composition of the Tunis tapestries echoes that of the Avalos series in the inclusion of handgunners in the foreground of several pieces. The work also provides evidence for perceptions of firearm use by the Ottoman Empire and its allies. Ottoman fighters are clearly shown with guns, reflecting their well-developed industry. That was not always the case in European depictions of conflicts between Christians and Muslims: two drawings by Vermeyen now in the Louvre show the latter with far fewer guns than the former.[52] In the case of Tunis, however, Vermeyen observed events directly (and included his self-portrait in the work). While his cartoons emphasized the numerical superiority of the Spanish infantry, including their arquebusiers, he also showed the Ottoman cavalry with guns, alongside some of their infantry (though a smaller proportion thereof than the Spanish) (Figure 8.8). A significant proportion of Ottoman fighters are depicted with bows. If the impression is of some technological advantage for Charles's troops, it is relative not absolute, reflecting broader contemporary analyses of Ottoman military prowess that might both over- and under-estimate their capacity.[53] Like the images of Mühlberg, the Tunis images were also adapted to other media: a plate now in the Louvre was manufactured in Antwerp in 1558–59 after Vermeyen's design.[54] Here again, they could make the journey from palace walls to more modest domestic settings.

Observation of battlefield practice is also apparent in Giorgio Vasari's wall paintings for the Sala dei Cinquecento—the great Hall of the Five Hundred in Florence's Palazzo Vecchio, by this point the principal residence of the Medici dukes. The two sequences, completed between 1567 and 1571, narrate conflicts over half a century apart: the Battle of Pisa (1499) and Battle of Marciano (1554), in the process illustrating the shifts in technology.[55] In the Pisa sequence, artillery is present in the composition, and crossbows are prominently featured, but there are no obvious handguns. In the Marciano painting, by contrast, crossbows have disappeared, and arquebuses appear in quantity. In the foreground, a pair of arquebusiers are shown, one aiming and one charging his gun with powder (Figure 8.9). Still, the pattern of no commanders with guns persists

FIGURE 8.8 Jan Cornelisz Vermeyen, *Tapestry Cartoons for the Conquest of Tunis by Charles V: An Unsuccessful Sortie by the Turks from La Goletta* (detail), 1546–50. Paper on canvas, charcoal and gouache, 385 × 832 cm. KHM, Gemäldegalerie, 2042. © KHM-Museumsverband.

FIGURE 8.9 Giorgio Vasari, *The Battle of Marciano*, 1567–71. Fresco, Palazzo Vecchio, Florence. Photo: World History Archive via Alamy.

from the Avalos tapestries: for all that we know the Italian elite owned and gifted guns, in the visual record firearm use remained the province of the ordinary soldier, not the men in charge.

Elsewhere in the Palazzo Vecchio's ducal apartments, in the Room of Leo X, handguns appear in images of both the 1517 conquest of San Leo (a fortress near Urbino, which assisted the Medici in seizing that duchy) and the 1521 recapture of Milan (where the city's defenders are shown shooting from the walls). Once again, these representations fit the established chronology for the advent of small arms into Italy. Given the involvement in the development of these decorative schemes of Van der Straet/Stradano—known for his interest in technology—this attention to detail should not come as a surprise. It is important to note, however, that personal tastes and interests mattered when it came to court decisions: the prominence of military technology and metallurgy in the decorative scheme for the Medici apartments in the Palazzo Vecchio in Florence is not matched by corresponding portrayals in the series showing the deeds of the Farnese in their residence at Caprarola (also painted in the 1560s). That, however, was a commission by a cardinal who sought to position himself as a 'worthy successor to his grandfather', Pope Paul III (r. 1534–49).[56] While the cardinal is shown accompanying Charles V and his troops to fight the Lutherans, handguns are notable by their absence from the image; taken as a whole the sequence emphasizes peacemaking and diplomacy over direct engagement in war. There are, moreover, numerous classical elements in the wider scheme, which along with the persistent idea of the handgun as ungodly would militate against guns' inclusion. The extent to which any given commission incorporated imagery of firearms thus depended on a number of factors, including the interests of the patron and artist, but over the course of the sixteenth century the repertoire of acceptable images slowly expanded.

The Battle of Lepanto

A case-study of particular interest is the iconography of the 1571 Battle of Lepanto.[57] This naval encounter saw an allied Christian fleet score an unexpected victory over the Ottomans, and in the context of the Counter-Reformation offered the Catholic powers an exceptional opportunity for propaganda. The continuing prospect of conflict with the Ottoman Empire had been one motivation for the maintenance of the Venetian arms industry. Indeed, since the Ottoman conquest of Constantinople in 1453, the Venetians had supplied its opponents with gunpowder weapons. Accounts of Giosafat/Iosafa Barbaro's voyage to Persia in 1471 describe him going with a large supply of weapons, including handguns.[58] During the reign of Uzun Hasan, sultan of the Aq Qoyunlu confederation from 1452 to 1478, the Venetians supplied the Safavids (rulers of Iran) with artillery.[59] None of this prevented the ongoing expansion of the Ottoman Empire both along the north African coast and through the Balkans, where in 1521 they captured Belgrade. There was further conflict with Venice in 1537–40, when the Ottomans made gains in what is now Greece, and the papal archives for this period show purchases of match-cord (and hemp for its production) as galleys were prepared.[60] Ottoman expansion continued through the 1560s and while the Ottoman siege of Malta in 1565 proved a failure, their conquest of the island of Chios from Genoa the following year proved a success.[61] This was the backdrop against which the Brescian arms industry prepared, in 1570, to produce a hundred thousand firearms a year. The Ottomans had a substantial firearm industry, which according to contemporary sources contributed to their success at the 1526 Battle of Mohács, and within a few years of that it seems that most of the elite Janissary forces were using firearms.[62] The depictions of military conquests in the *Süleyman-name*, commissioned by the Emperor Süleyman the Magnificent as a history of his reign and completed in 1558, show handguns in use, for example, during the Siege of Rhodes in 1522.[63] The ambivalence about firearms further west does not appear to have had a counterpart in the Ottoman Empire.[64]

When showed at panoramic scale, as was common, naval battles provided fewer opportunities for the detail of firearms provided in the earlier tapestries and wall-paintings of land wars. The best detail appears in Vicentino's paintings for the Sala dello Scrutinio of the Palazzo Ducale, Venice, but there is no equivalent of the close-up views of loading found in the Avalos tapestries or Palazzo Vecchio frescoes. Martin Rota's engraving shows some detail of handguns but the clouds of smoke predominate.[65] No doubt this in part reflects a desire to show the scale of the naval encounter, which required firepower to be represented by clouds of smoke rather than individual weapons (as in Vasari's depiction of Lepanto for the Sala Regia in the Vatican or the anonymous painting now in the Museo Correr). As Christina Strunck has argued, however, the Sala Regia fresco does point in several ways towards 'Turkish inferiority', including in relation to weaponry: the Christians here use firearms, while the Turks do not.[66] Titian's *Philip II Offering the Infante Fernando to Victory* clearly shows a background battle with gunpowder weapons, while a captured Ottoman fighter is shown with arrows, like Vasari suggesting some technological disadvantage.[67] The same absence of firearms is apparent in the anonymous portrait *The Victors of the Battle of Lepanto* now in the Kunsthistorisches Museum, Vienna, although that also reflects the general rarity of firearms in sixteenth-century portraiture.[68] The imagery of Lepanto was further shaped by the broader shifts in Counter-Reformation art already noted. Veronese's painting of the battle, now in the Accademia, Venice, shows saints looking down onto the navies, a composition for which obvious inclusion of firearms may have been judged inappropriate.

Guns were certainly an uncomfortable fit into literary narratives of the Battle of Lepanto, stressing as they often did the Christian victory, which took place at a time shortly after the attempt on Carlo Borromeo's life and the shooting of the earl of Moray, when the popes were cracking down on the 'impious' weapons. Describing the battle in his 'Hymn to St Mark and St Justina', Davide Podavini described how 'the heavens rumble with the crash of arms. The stunned earth trembles and the hollow cliffs resound. Shores as far away as India groan, and the burning dust envelops the sun and the sky, and the air is choked by unfamiliar clouds.'[69] Agostino Fortunio likewise wrote of 'clouds of gunfire and smoke', and Giovanni Battista Amalteo described how Sebastiano Venier, admiral of the Venetian fleet (to whose family Giovanni Battista Porcellaga had gifted weapons), was sending 'avenging ships and bronze thunderbolts against the enemy; the waves trembled at the sound of gunfire bursting from the cannons'.[70]

The most striking of the various accounts for its explicit reference to handguns was Guglielmo Moizio's 'Song on the Victory of the Christian Fleet'. Even cannon, which tend to receive a more positive portrayal in these poems than handguns, were for Moizio a 'horrible sight'.[71] As for handguns:

> On both sides the missiles fly. So many barrels, made either from hollow iron or from flashing bronze, rain with sulfur, and, ignited by gunpowder and ruddy flame, spread bullets of blue-black lead flying through the air like hail; struck by these, many thousands of men fall. This is what teachers meant by "scloppum", the blast of air that came from Stygian chimneys.[72]

The reference here to the River Styx reinforces the connection of guns to Hell. Moizio was particularly critical of the practice of shooting from a hidden position: 'Concealed from fire and iron, the soldiers scatter wounds into the dense crowd from afar, sending bullets and cruelly shrieking lead through the air, and unseen—shamefully—they send brave bodies to death.'[73] His poem continued with an extended diatribe against 'detestable gunfire' and guns, 'manageable even in the hands of the despicable and weak soldier'. The lethality of firearms was a concern too, not least for the 'poor wretches, on whom gunfire inflicts untreatable wounds as it breaks through the ribs with burning lead'.[74] It is a sharp contrast with the portrayal shortly afterwards of 'the noble youth of Italy . . . trained for extended wars in close quarters and accustomed to fighting with the sword'.[75] This rhetoric shows the persistence of the narratives established earlier in the century by writers like Ariosto. Guns might be war-winning weapons, but they remained unchivalrous.

It would be easy to assume that the Battle of Lepanto furnished a straightforward opportunity to situate firearms in the Christian arsenal. Yet although works by Vasari and Titian hint at Christian technological advantage, this narrative was by no means the standard in either the literary or the visual sources. One problem, of course, was that the Turks had firearms too, and in different circumstances the gun-given victory might have lain elsewhere. The continuing diabolical associations of guns made their incorporation into the Catholic iconography awkward, especially where the representations included images of saints. Given the many factors involved, from the prior literary tropes to the contradictory attitudes towards the Ottomans, to the new demands of Counter-Reformation art, some messiness is only to be expected.

Religious wars

If firearms were sometimes an awkward fit in the heroic narratives of the Battle of Lepanto, however, they were highly present in the visual culture of both the French Wars of Religion (1562–98) and the Dutch Revolt (1566–1648). Sometimes in these contexts the portrayals deployed the existing negative tropes of the firearm. Yet by this time guns were common enough that they routinely appeared in any imagery aiming to depict recent events. Thus François Dubois's painting of the St Bartholomew's Day Massacre in 1572, one of the very few contemporary images of the event, features multiple figures bearing firearms.[76] There is a similarly prominent presence of handguns in Frans Hogenberg's print of the 1588 killing of the Guise brothers, produced between 1589 and 1593.[77] Yet the representations are not always straightforward. Tom Hamilton has argued that the multiple paintings showing the well-armed 1590 Procession of the Catholic League in Paris are in fact satirical.[78] Rather, they mocked the incompetence of the marchers, especially

FIGURE 8.10 Unknown artist, *Allegory of the Church* (*Pièce allégorique sur L'Église*), 1569–76. Engraving. Bibliothèque Nationale de France / Gallica. Public domain.

the clergy toting unfamiliar weapons. They drew attention to the story (confirmed in textual sources) of one monk who accidentally shot and killed a bystander, possibly the secretary of the papal legate, the incongruous combination of guns and priests adding to the satirical effect.

The visual culture of the Dutch Revolt likewise drew on well-established cultural associations of firearms, both positive and negative. A print of c. 1572 by Theodoor de Bry contrasted the Protestant William of Orange with the Catholic Duke of Alva. Titled *Vices Rule the World; Truth and Justice Sleep*, it showed Alva accompanied by figures representing vice, including Lust, who is firing a gun (pistols had also become a sexual metaphor). There were no guns on the side of truth and justice, whom William was trying to awaken.[79] The engraving *Allegory of the Church* (Figure 8.10) showed worshippers fleeing naked and unarmed as the Church of Christ came under attack from the Antichrist and his household or soldiers, who had handguns; from the Turks; and from the duke of Alba, the cardinal of Lorraine and Cardinal Granvelle, who were firing cannon.[80] Here, it was those on the side of the Devil who used guns.

On the other hand, military service in the course of the Dutch Revolt, combined with the prior acceptance of guns in the iconography of the civic militia, offered new possibilities for portraiture. A 1567 painting by a follower of Anthonis Mor van Dashorst shows Philip III von Croÿ, duke of Aarschot, his right hand gripping the upright barrel of a handgun.[81] The young Maurice, prince of Orange, was painted around 1580, at the age of about twelve or thirteen, holding a small wheellock pistol.[82] This trend may have influenced English painting of the same period, when we find portraits of Sir Thomas Tresham (1568), Sir William Drury (c. 1570–75, attributed to George Gower), Robert Dudley (c. 1585–86, attributed to William Segar) and William Stanley (who initially fought alongside Dudley in the Dutch Revolt), all shown with guns.[83] Dudley's portrait is of particular interest because following the assassination of William the Silent he led English troops in the Netherlands between 1585 and 1587. This was one of a number of works he commissioned (Figure 8.11).[84] The army camp is shown in the background, while Dudley poses with his left hand on a helmet and his right on a petronel, a type of firearm with a distinctive curved stock. Elizabeth Goldring, author of the principal study of Dudley's art commissions, suggests that the portraits may have been inspired by those of Dudley's opposite number in the Dutch campaign, Alessandro Farnese, duke of Parma.[85] The gun, however, does not find a parallel in paintings of the Italian, which may reflect the differing cultural expectations of Italian portraiture. Lisa Jardine notes that the portrait shows Dudley after the 'ignominious end' of the campaign, suggesting that the hand over the gun's muzzle is a 'symbolic acknowledgement of lost virility'.[86] It is certainly not the image of a man about to leap into action. It also, however, marks an alternative to the representation of guns in the De Bry engraving: if there they belong to the entourage of the Antichrist, here they are the elegant weapon of the Protestant commander even if, on this occasion, he is not the victor.

A later design by Stradano for *Scheme or Mirror of Princes* offers further evidence for the malleability of firearm iconography towards the close of the sixteenth century. Printed after 1594, this

FIGURE 8.11 Unknown artist, *Portrait of Robert Dudley*, 16th century. Parham House, West Sussex.

gave a more balanced view of firearms across its five engravings and a frontispiece as it set out the duties of the ideal prince.[87] Taking inspiration from the *Allegory of Peace* produced by Francesco Salviati for the Sala dell'Udienza in the Palazzo Vecchio, in one engraving a female figure sits atop a pile of weapons, her foot on a handgun. This implicit rejection of the weapon, however, was complemented by a positive representation in the image devoted to Hunting. If guns were still to be treated with caution in some regards, they undeniably had acceptable uses.

Empire and guns in portraiture

Netherlandish artists were more broadly important in contributing to a series of English portraits that in the later part of the sixteenth century showed men with guns. All of them, moreover, have a link to religious conquest. The first is that of Thomas Butler, tenth earl of Ormond. Attributed to the Netherlandish artist Steven van der Meulen (active 1543–68), it shows Butler holding a wheellock firearm (Figure 8.12). The use of such guns was heavily regulated in England, but Butler was a distant cousin of Elizabeth I (on her mother's side). He had been brought up as a Protestant at the English court, and appointed Lord Treasurer of Ireland from 1559 (at Elizabeth's accession) until his death in 1614, during which period he was a central figure in the Tudor conquest and colonization process, through the Desmond Rebellions of 1569–73 and 1579–83 as well as the Nine Years War of 1594–1603 which together paved the way for the seventeenth-century Plantation of Ulster. The unusual iconography can most straightforwardly be explained by the fact of Butler's military role in a frontier environment. The 1594 portrait of Captain Thomas Lee by Marcus Gheeraerts the Younger is more iconographically complex, but its subject likewise has a gun, in this case an inlaid snaphaunce pistol. Lee (1552/53–1601) was involved in the English campaign in Ireland, to which the portrait makes reference via his depiction bare-legged as a common foot-soldier; a quotation from Livy alludes to the Roman hero Caius Mucius Scaevola, whom we encountered in chapter 4.[88] Here the firearm is integrated into a learned elite culture in which viewers of art could be expected to recognize the classical reference, once

FIGURE 8.12 Attr. Steven van der Meulen, *Portrait of Thomas Butler, 10th Earl of Ormonde*, mid-16th century. Oil on panel, 93 × 68 cm. National Gallery of Ireland, NGI.4687. © National Gallery of Ireland.

again quite some distance from a literary culture that had continued to be hostile towards firearms.

Imperial and colonial projects had broader implications for the representation of firearms across Europe. A 1577 portrait of Sir Martin Frobisher by Cornelis Ketel shows its subject holding a pistol; behind him on a table sits a globe (Figure 8.13). Ketel, who was from the Netherlands, may have shared his homeland's more relaxed attitude towards the depiction of firearms in portraiture: he went on to paint a company of the Amsterdam militia, duly armed with guns.[89] Even so, the image of Frobisher is unique. Frobisher had joined the first English expedition to West Africa while still in his teens; on a second expedition he was captured by the Portuguese and imprisoned; he was, however, most famous for his three voyages made in search of the North-West Passage. As a single image conveying the significance of the gun in narratives of conquest, Ketel's is hard to beat. It was originally one of a series of fifteen paintings, the others featuring various of Frobisher's colleagues and three Inuit people captured during the voyage.[90] The imagery of the gun in hand exemplifies the 'weapon in hand' described by Gonzalo Fernández de Oviedo y Valdés in his *Natural and General History of the Indies*; Oviedo referred to Spanish colonists as always 'con l'arme in mano'.[91] (Whether the colonists necessarily had small guns at the point Oviedo made his observation is another question.) Readers who encountered Oviedo via the compilations made by Giovanni Battista Ramusio, whose *Navigations and Voyages* were certainly read in England, would have gained a corresponding image from multiple tales of Indigenous peoples scared by firearms, to which we will come in chapter 9.[92]

Of the surviving sixteenth-century English portraits of men with guns, nearly all have a connection to religious war or colonial conflict (and of course, the two were interlinked). They are also primarily the work of Netherlandish artists, who may have been influenced by the distinct culture of civic gun use in the Low Countries. Maybe there are other, lost, portraits: even so, it seems intriguing that these were kept. While Lois Schwoerer does not neglect to mention Ireland or the Cathay company in her consideration of the Lee and Frobisher portraits, the thematic connection between the two is lost in her wider discussion, which focuses more generally

FIGURE 8.13 Cornelis Ketel, *Portrait of Sir Martin Frobisher*, 1577. Oil on canvas, 211 × 98 cm. Bodleian Libraries, LP 50.

on martial and aristocratic masculinity.[93] Rather than seeing these portraits primarily as evidence for a growing acceptance of handguns among the English elite, there is a strong case for seeing that acceptance as conditioned by the particular circumstances of their use—that is, outside the parameters of conventional war between European princes.

Conclusion

While the later fifteenth century saw a number of inventive and experimental representations of firearms, the literary hostility towards guns in the sixteenth century, along with the preference of some patrons especially in Italy for classicizing styles, shaped the environment in which firearms were depicted. Bar a handful of exceptions, noble portraiture avoided the inclusion of firearms. Artists who wished to convey the sitter's engagement with martial culture typically did so via armorial imagery evoking the chivalrous knight and tournament. When firearms began to appear in portraits of the European royalty and nobility, as we have seen in the examples from Velázquez, they were more generally associated with the hunt. Battle scenes such as those of Vermeyen, which aimed for a sense of documentary reality, certainly included guns, illustrating the contrast between the two sides in their usage of this technology; Vasari's frescoes for the Palazzo Vecchio likewise conveyed a sense of technological development. Yet while Christian art of this period frequently showed biblical figures in modern dress, and with modern weapons, the depiction of guns in this context was vanishingly rare; in the fifteenth century a few artists showed crossbows, but by and large this experimentation did not last. Not until the seventeenth century, when cases like that of Carlo Borromeo provided reason to show shooting in relation to saints' lives, did things change. For biblical scenes, the particular expectations of Counter-Reformation art mattered: the exception to the rule was provided by an artist working outside the usual process of commissions for church interiors. Bruegel's Low Countries environment had a more open attitude towards gun use, as illustrated by the early group portraits of the Amsterdam militia. Netherlandish artists working in England were responsible for further portraits featuring firearms:

FIGURE 8.14 Bartholomeus van der Helst, *Militia Company of District VIII in Amsterdam under the Command of Captain Roelof Bicker* (detail), Amsterdam, c. 1640–43. Oil on canvas, 235×750 cm. Rijksmuseum, SK-C-375. Public domain.

there was also a notable trend here towards showing guns in imperial and colonial contexts. With few exceptions, the acceptable contexts to be shown with a gun were the regulated city militia, or the frontier.

A seventeenth-century painting of the Amsterdam militia brings these two contexts together in striking fashion. The most famous work of this period is Rembrandt's *The Night Watch*. Figure 8.14 shows a close contemporary daytime parade. Painted in 1643 by Bartholomeus van der Helst, it shows the Militia Company of District VIII under the command of Captain Roelof Bicker.[94] In terms of firearms and race, here we have the polar opposite of the Tournai tapestry. There we saw a single Black figure with a handgun. Here we see a single Black figure without one, while in contrast all the white members of the militia are carrying guns. The Black boy turns his head nervously towards the gunman behind, whose weapon is equipped with a lighted match and raised ready to fire. In contrast, a white boy looks confidently out from the canvas, polearm in hand, unperturbed by the presence of the lighted match to his rear, perhaps because that gun is clearly not being aimed. The Rijksmuseum curators hypothesize that the Black boy may have been the servant of Bicker (the figure to his right); living in Amsterdam he must legally have been free, although in practice he may not have known that. Over two centuries the association between guns and race had flipped. In the 1460s, a firearm could be portrayed in the hands of a Black figure, a depiction most likely drawing on the negative associations of the two. As European powers expanded their empires, the gun, while still suspected, acquired a new association with European conquest and with European technological superiority.

CHAPTER NINE

The European Empires

THE BURGKMAIR DESIGNS for the Triumph of the Holy Roman Emperor Maximilian, produced between 1512 and 1519, tell us not only about attitudes to firearms within Europe, but also about weapons beyond the continent. Several sections show the peoples of Calicut (Kozhikode, India, but here probably used in a looser sense) carrying bows, arrows or spears, a sharp contrast with the European troops elsewhere in the sequence who are equipped with firearms, including handguns. (Figures 9.1a–c and 1.9.) The contrast is misleading: gunpowder technologies were available across Asia, and by the time the designs were completed, Java, for example, could boast of 'great masters in casting artillery; they make many spingards [a type of artillery], *schioppi* and fireworks, and are everywhere reputed excellent in this craft of casting artillery and in knowing how to fire it'.[1] Yet despite the reality, the visual contrast served a purpose in conveying the superiority of European military technology to audiences across the Holy Roman Empire. Europe's firearm revolution, the story went, was enabling its rulers to conquer the world.

The Christian powers were not supposed to supply firearms to non-Christians, although Venice did so, somewhat on the basis that the enemy of my enemy is my friend, shipping weapons not only to the Mamluks but also to the Aq Qoyunlu (rulers until 1501 of what is now eastern Turkey, Armenia, Azerbaijan and parts of Iraq and Iran).[2] Guns found their way to India via the Portuguese, who like the Venetians were prepared to ignore the papal ban on weapons sales to non-Christians, and used them as diplomatic gifts.[3] The

FIGURE 9.1A–C People of Calicut, from Hans Burgkmair, *Triumph of the Emperor Maximilian I*, 1512–19, Augsburg. Woodcut on paper, approx. 38×38 cm. Victoria and Albert Museum, 13079.119. © Victoria and Albert Museum.

improved European weapons technology also made it full circle back to China, where the innovations were adopted, and to Japan.[4] As in the case of Francis I at Pavia, there is some evidence that dishonourable firearms were blamed in literary culture for the fall of the chivalrous troops of the Mamluk Empire (which was conquered by the Ottomans in 1517). Mamluk historian Ibn Zunbul had his imagined Mamluk officers account for their loss with reference to firearms. As Robert Irwin argues, however, this should not be taken as a literal account of the wider culture, just as the French tales of

FIGURE 9.1B (*continued*)

Francis's chivalry should not be assumed to represent the sum of courtly attitudes.[5]

The precise role of gunpowder technology in the development of the European empires is still not fully understood. Firearms were a minority of the weapons used in conquest, and historians now rightly avoid simplistic narratives of technological superiority, emphasizing factors such as alliances with Indigenous rulers.[6] Yet for all that firearms came to prominence precisely in tandem with the European empires, for their earliest period there is rather limited close analysis of what the sources do tell us about the relationship between the two, though new work by Hyeok Hweon Kang is making an important contribution, especially for later

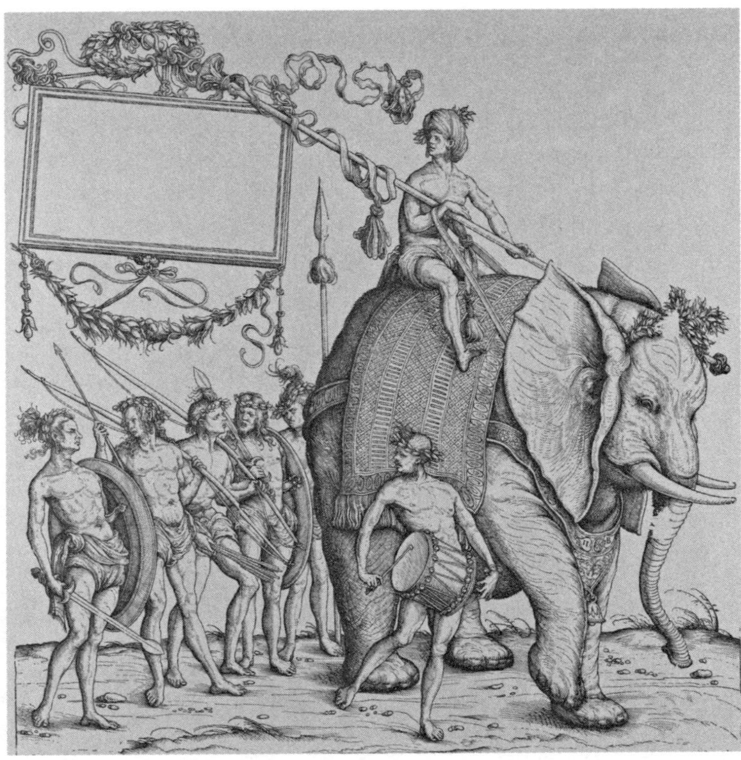

FIGURE 9.1C (*continued*)

sixteenth-century Korea.[7] For the seventeenth and eighteenth centuries, there is relatively more historiography, notably David J. Silverman on the Native American acquisition of firearms, Priya Satia on guns in the British Empire, and Toby Green's work on the economy of precolonial West Africa, where there were important intersections between the gun trade and enslavement.[8] The sources for the earlier period are more fragmentary, but by exploring both literary and artistic production it is certainly possible to see how within Europe a narrative highlighting the advantages of handguns in conquest became established. Beyond the initial shock value of firearms, it is in fact questionable how significant they were (a point not lost on early authors). Nor, as Frank Jacob and Gilmar Visoni-Alonzo have pointed out, were wider western European military innovations of great assistance in imperial expansion.[9] My interest here, however, is not so much in what was really happening outside Europe, but in how the stories that were told about conquest informed wider European attitudes to guns. Imperial expansion was justified by prospects for conversion to Christianity, and these narratives offered an important counterweight to the idea that the gun was ungodly.

This is a much larger topic than can be contained in a single chapter, of course, so in line with the geographical focus of this book on Italy I have taken as case-studies those accounts collected in Giovanni Battista Ramusio's *Navigations and Voyages*—published in Venice and one of the most influential texts on European understandings of the wider world—and Book 12 of the *Florentine Codex* compiled by Fray Bernardino de Sahagún and his Indigenous collaborators, which drew on the oral accounts of men present during conquest of Mexico and informed artistic production at the Medici court in the later sixteenth century, notably the ceiling paintings of the Armoury in what is now the Uffizi Gallery, Florence. Finally, the chapter considers representations of guns by artists in early modern West Africa, Peru and Bolivia, showing that they too perceived firearms to be a significant element in European culture.

Morocco

As we saw in chapter 8, representations of Ottoman weapons in Europe sometimes, but not always, implied some technological inferiority. An early version of this trend is apparent in narratives of conflict between Portugal and Morocco. Weston F. Cook has explored the significance of gunpowder in Moroccan state formation over a series of wars from the 1460s to the late sixteenth century. These were complex conflicts, involving Iberian incursions and a civil war between rival ruling families, as well as local revolts. At first, firepower gave the Portuguese an advantage in conquering key ports and territory, although as the Pastrana tapestries show, their opponents acquired gunpowder weapons early. Over the course of the sixteenth century, the Moroccans successfully resisted both Iberian and Ottoman conquest and in 1578 beat the Portuguese at the Battle of Wādī-al-Makhāzin.[10] In 1591 a Moroccan force defeated the Songhai Empire (which did not have gunpowder weapons) at the Battle of Tondibi, leading to that empire's collapse.[11] Iberian sources for the early stages of the Moroccan conflicts emphasize the psychological impact of guns. Hernando del Pulgar, discussing the siege of Sentenil in 1484, wrote: 'Such was the terror the firearms inspired and the carnage and ruin inflicted upon the Moors they could not endure it.'[12] Describing an invasion in 1508, Aragonese court chronicler Jerónimo Zurita y Castro noted: 'The King of Fez, Muhammad ash-Shaykh, had sent a great many forces to the qā'id to oppose the landing of the Christians. But these contingents, out of fear of the Duke's shipboard cannons, let the city fall.'[13] Other writers place this access to technology in a wider context of European superiority. Duarte Pacheco Pereira, a Portuguese captain and cartographer, wrote (also in 1508): 'Pliny says that Europe, being more excellent than all the other parts of the world, produces more conquering races . . . nor can those who live in Asia or Africa deny that Europe possesses great abundance of arms and so much artillery'.[14] These were partisan sources, however, deploying what would soon become a trope.

Al-Hasan ibn Muhammad al-Wazzan, known to his European readers as Leo Africanus, whose family was from Fez, observed in his *Description of Africa* (written around 1526 although not published until the 1550s) that firearms were used in conflicts between local rulers in the Sūs and Dra'a areas of Morocco: 'The people of this country have been using arquebuses and *schioppetti* for some time. I have never seen better shooters: they would hit the tip of a needle. With these arquebusiers they massacred each other.'[15] This suggests a significant proliferation of firearms into Moroccan territory in the first decades of the sixteenth century. Indeed, a late sixteenth-century French account of Morocco suggests that gunpowder technologies might be viewed in positive terms, indeed emphasising their gallantry. At the Third Battle of Rukn in 1596, Crown Prince ash-Shaykh al-Mā'mūn, 'seeing his frontline troops heavily engaged, dismounted among his gallant musketeers who crowded around him. Taking a musket in hand and the shot in his mouth, he called out: "At them, brothers and comrades, now I'm going to die with you or we win."'[16] The observation, however, reflects as much shifting attitudes towards firearms within Europe—by the end of the century musketeers were permitted to be gallant—as the facts of the Moroccan conflicts.

The Horn of Africa and Asia

The *Navigations and Voyages* collected and published by Giovanni Battista Ramusio between 1550 and 1558 were among the most important sources for European knowledge of other continents. Over the course of several editions, the collection's coverage came to span the world from present-day Myanmar and Indonesia via the Indian subcontinent and Africa to the Americas, also covering the Safavid and Ottoman Empires, Eastern Europe, including Poland-Lithuania and Muscovy, and to the far north, Greenland. Printed in Venice, a major centre for distribution of news about European expansion, it was fundamental to subsequent scholarship, and parts were translated, including by Richard Hakluyt, into English.[17] Ramusio's diplomatic connections enabled him to include many documents that had not been published elsewhere, and while some sections of the text have now been identified as fake (including the Vespucci letters and the account of the non-existent Frislandia), much of it reflects the genuine, if biased and confused, perceptions of travellers and diplomats.

A reader of Ramusio would be well aware that while the earliest voyage narratives pointed to relative ignorance of guns, rulers in East Africa, India and beyond quickly acquired gunpowder technologies and became adept in using them. An account of an anonymous 'pilot' (more likely a clerk) accompanying Pedro Alvares on the voyage of 1500 most notable for Alvares's landing in and claiming of Brazil described how the crew made a gun salute in Calicut, 'at which they [the local people] marvelled greatly, saying that no one but God could overpower us'.[18] This is an early example of what would become a well-worn trope in European stories of encounter (indeed, we have already noted its use in relation to Morocco). When Alvares and his men landed in Melinda (Malindi, Kenya), they did so (on Alvares's account) carrying concealed weapons.[19] Concealment was not generally the strategy of Europeans, who more commonly wore weapons openly, but two years earlier the ruler of Malindi had been welcoming to Vasco da Gama (a potential ally against a rival in Mombasa), so this was most likely a means of indicating friendship. Alvares does not record the type of weapon used here, and at this early stage it is unlikely to have been a firearm, however, the incident illustrates a scenario in which concealable guns might also have been attractive. More generally, gunpowder weapons soon proliferated. Lodovico de Varthema of Bologna, whose account of his travels via India to the Spice Islands was published on his return in 1510, observed artillery in Cevul (perhaps Kelve) in Gujarat and Canonor (Kannur); he saw an armada from Calicut with 'artillery large and small after our fashion'.[20] Duarte Barbosa, a Portuguese official in Kannur whose father and uncle had also acted as agents in India, composed his account of travel to Gujarat in 1517–18.[21] He described *schioppetti* being fired from the top of battle elephants and firearms being used in Brahmin and Lingayat ceremonies.[22] The ceremonial use of guns by European navies was quickly established in these early encounters, and it must fairly swiftly have been understood that it was not necessarily hostile, even if it did underline the presence of potentially destructive technology.[23]

On their voyages, the European navigators made presents of guns, just as they did to fellow rulers on their home continent. Antonio Pigafetta, one of the handful of men to survive the circumnavigation initiated by Magellan (1519–22), described in his chronicle of the voyage a gift made to the king of Tidore (one of the Malaku islands, Indonesia), where the ships took on water. It consisted of 'some pieces of artillery, small like arquebuses' along with 'four barrels of powder and some glass beakers'.[24] The king of the nearby island of Gilolo (Halmahera) came on board ship and saw the weapons store, including *schioppi*, crossbows and large *verzino* (brazilwood) bows; Pigafetta reported that he preferred shooting with crossbows over *schioppi*, but gave no explanation as to why this was the case.[25]

Gunpowder technology was also presented to Dawit II, king of Ethiopia, by a Portuguese embassy in 1520–26. The Portuguese sources suggest this was a novelty, but firearms had certainly been used in the Mamluk Empire and Leo Africanus refers to the gift of a handgun around 1512 by an Egyptian traveller to the king of Gaoga (in central Sudan), which does raise some doubt about just how new guns were.[26] Ethiopia was unique in Africa at this time as a Christian power, and the Portuguese cast Dawit in the role of the legendary Prester John, lost Christian ruler of the East. An account of the mission by Francesco Alvarez, the ambassador, was first published in Lisbon in 1540 before being translated in Ramusio's collection, and included a list of the diplomatic gifts:

> A sword and dagger, very rich and beautiful; four arras cloths with figures to cover the walls, very fine; a beautiful cuirass [upper body armour] covered in velvet and a rich gilded *celatone* [large sallet-type helmet], two pieces of artillery with four tails [*code*, to support the gun], some cannonballs (*ballotte*) and some barrels of powder, a mappamundi and an organ.[27]

Alvarez narrated how the response to the embassy's firearms shifted from suspicion to enthusiasm. Asked initially why they had not brought many spingards up from the sea, the ambassadors replied that they were not there to make war but had brought up three or four 'for fun and to pass the time', firing them with powder but not shot to demonstrate. There followed a conversation between Dawit and the ambassador about the Turks and their firearms, after which Dawit enquired whether the Portuguese could help with making gunpowder and training troops. Evidently, he saw an opportunity to press for a Christian alliance against a common Muslim rival. The ambassador later told Dawit that the gifts they brought were for demonstration, and that he should let them know how many he wanted.[28] Once again, this echoes practices of giving guns at the European courts. In a letter of 1524 to King João III of Portugal, Dawit requested that he send artisans skilled in gunmaking: 'all artisans will be very dear to me, and very much what I need, especially those who know how to make *schioppetti*'.[29] The reference to the smallest of small arms here hints at the particular difficulties associated with their reliable production. On Alvarez's account, the only artillery the Ethiopians had were fourteen spingards purchased from Turkish merchants; they had no bombards except for the two tails brought by the embassy itself.[30] (The inconsistency of this description with the earlier one of two artillery pieces with four tails is not explained.) A later English account of *The Great Princes of Africa*, published in a volume with Leo Africanus's description of that continent, described how by the 1560s the Portuguese were actively arming allies in the country: they 'do daily bring into Abassia

FIGURE 9.2 Unknown maker, *Continuation of the Triumphal Procession: Presentation of War Material*, Brussels, 1555–60. Tapestry: wool, silk, gold and silver, 350 × 326 cm. KHM, Kunstkammer, T XXII 8. © KHM-Museumsverband.

[Ethiopia], the manner of warfare in Europe, with our use of armes, and the manner of fortifying passages and places of importance'.[31] The visual sources of mid-century reflect these narratives, showing the large quantities of European weapons that were shipped east. Particularly notable are the tapestries woven in Brussels to commemorate Dom João de Castro's 1547 conquest of Goa, one of which illustrates the war materials with matchlock firearms prominent in the foreground (Figure 9.2). What these tapestries do not, however, show, is that the war materials were needed because Goa already had a well-developed local firearm industry.[32]

None of this is to say that firearms invariably protected Europeans, nor that the narratives portrayed them as doing so. Pigafetta's account of Magellan's death in 1521 in Mathan (Mactan, the Philippines) is a case in point. Having failed to persuade Lapulapu (Silapulapu), the island's ruler, to make his obeisance to the king of Spain, Magellan's men faced off against a thousands-strong army. Lapulapu's men stayed out of range of their gunfire while the crew fired constantly for half an hour, thus (we may infer) exhausting the accessible powder and shot, at which point Lapulapu's troops attacked.[33] This is as much a cautionary tale about poor tactical use of firearms as about firearms themselves, but it does confirm their limitations. Still, the overall picture conveyed by Ramusio's collection in relation to the Horn of Africa, India, Indonesia and beyond is that while access to firearms and powder had initially been very much in the gift of the European powers, used as a bargaining tool to win friends and intimidate enemies, already in the first quarter of the sixteenth century the technology was widely diffused and being locally produced. By 1569, Cesare de' Fedrici (whose account appeared in the 1587 edition of Ramusio) was reporting on local arquebusiers drilling daily in Pegu (Bago, Myanmar) and also claiming that elephant skin could repel shot.[34] Cesare, a jeweller, was originally from Valcamonica, one of the Brescian valleys, and would have been familiar with arms and armour. Perhaps he had taken a gun from his hometown with him.

Southern and Western Africa

In southern Africa, on the other hand, the picture of firearms conveyed in Ramusio's collection was different. Reaching the coast of Mozambique (after a stop on the coast of what turned out to be Brazil), the Alvares crewmember reported that gunfire from the ship scared the local residents, the familiar trope once again.[35] Giovanni da Empoli, a Florentine merchant travelling with the Portuguese, explained that at Mossel Bay (South Africa, just past the Cape of Good Hope) 'the men carry certain darts with an iron tip'.[36] As we will see, this distinguished them from some Indigenous peoples in the Americas, who did not have iron. Giovanni's description of the African mainland, to which he travelled from

Mozambique, provides an example of how Europeans perceived guns in the context of these early encounters:

> Sometimes we used to go at our pleasure to the mainland, to see the countryside, where we found several generations of people all black and all naked. . . . They are very timid, especially when armed men come: seeing these beasts to be few and wretched, we gathered together about five or six companions, very well armed with *schioppi*, and took a guide from the said island who led us across to the country, and we spent a good day on the mainland.[37]

Guns thus offered the Europeans the ability to intimidate the people they encountered, and to enjoy their explorations with a feeling of security. We might question why such arms were required in the face of the 'few and wretched' people Giovanni describes, but the inconsistency emphasizes that the gun as narrated here is also an indicator of European superiority.

The emphasis on guns prompting fear persisted into later European accounts of West Africa. Pieter de Marees, perhaps a Flemish migrant to the Netherlands, made several trips to the West African coast, although the reasons for these are not known.[38] He gave an account of the Portuguese in 1570 burning the towns of Comando and Fetu, then themselves coming under attack, with losses running into the hundreds as they retreated to their fortress (Elmina, in what is now Ghana). The Africans, he explained, 'developed such antipathy towards the people of the Castle that they would have taken the Castle from them if they had not been afraid of their Artillery'.[39] His subsequent observations, however, paint a more nuanced picture of attitudes towards guns:

> Although they [the Africans] are greatly afraid of the shooting of big Guns or Muskets, nevertheless they enjoy hearing gunfire; for, when Ships come to anchor or sail away, they come out of their Huts and run towards the beach in order to hear the salutes of the Ships. When Traders come to the Ships and spend a fair sum of money, they desire that a Gotelingh be fired in their honour. They also buy many Firelocks and are beginning to learn to handle them very well. They have understood that a long Gun carries further than a short one, as the Portuguese, and we too, sufficiently teach them.[40]

In a later section of his account, he explained that the force of 'over six or seven hundred black or Mulatto Slaves who are their subjects and under their [Portuguese] orders know quite well how to handle weapons (Guns as well as Spears)'.[41]

In short, much as they had in Asia, the Portuguese had proceeded by arming and training allies. Guns suited West African practices of infantry warfare in forest territory, although the climate with its high humidity meant they had to be carefully maintained.[42] That may be why the traveller Samuel Brun, observing the coast on several voyages during the 1610s, noted that the men sent by the kings of Fetu and Sabou to help with the fight against the Spanish had guns that were 'polished beautifully clean'.[43] Officially, the Portuguese prohibited the sale of guns to West Africans, but pragmatism required that their local allies be armed, and gradually other European powers likewise began selling arms into the region. During the period of the transatlantic trade in enslaved Africans, the export of gunpowder to the region became a lucrative business.[44]

The Americas

Returning to Ramusio, similar narratives are to be found in the collection of documents relating to the Americas, although it is notable that descriptions of the early conquests do not refer to any arming of allies. The portrayal of guns in these texts clusters around several themes, of which the Indigenous lack of effective weapons is one (the exception, in some cases, was poison darts, which did concern the European writers). Another is the fear that might be prompted by artillery and firearms, although, as Camilla Townsend points out, Aztec warriors in fact provided quite dispassionate analysis of the new technologies.[45] The European writers were also conscious, however, of some of the limits of guns, particularly in relation to the need to secure supplies of powder. Most importantly, however, in terms of how firearms were understood in Europe, the stories began to create an image of the gun as a godly instrument, and did so before Europe's own mid-century religious conflicts.

The earliest accounts of voyages to the Americas include a good deal more discussion of artillery than handguns, and it is clear that only a small minority of soldiers carried arquebuses and *schioppi*. Michele da Cuneo, who travelled to Jamaica in the company of Christopher Columbus, described how their boats were equipped 'with shields, crossbows and bombards'.[46] Pietro Martire d'Anghiera, an Italian royal tutor, diplomat and historian at the Spanish court, where he had access to contemporary records of and correspondence concerning the conquests to inform his *On the New World*, published in Venice in 1534. Martire explained in general terms how when Bartolomeo Colombo left for the gold mines in late 1495 or early 1496, he went with 'some well-armed men', leaving it unclear whether in fact they used guns.[47] He was similarly unspecific in recording the 'great quantity of artillery and other arms' being taken by Martín Fernández de Enciso when he left for Uraba around 1509–10;[48] he described Vicente Yáñez Pinzón using artillery to intimidate on his 1508 arrival in Cuba.[49] Gonzalo Fernández de Oviedo y Valdés's *Natural and General History of the Indies*, influential in its own right and partially incorporated into Ramusio's volumes along with his shorter *Summary*, was like Pietro Martire for the most part unspecific about the detail.[50]

Even during the conquest of Mexico, firearms were used in significantly smaller proportions than in contemporary wars within Europe.[51] The relations of Hernán Cortés make clear that only a minority of his troops had handguns. In his third relation, dispatched in 1521, he reported how he had sent to Hispaniola for further arms; a muster at Christmas 1520 in the city of Tascaltecal turned out 'forty cavalry and 550 infantry, of whom eighty were using crossbows and *schioppetti*'; Cortés had eight or nine pieces of field artillery 'and a little powder'.[52] Subsequently he described a force of two hundred Spaniards, 'of whom there were eighteen on horseback, thirty with crossbows and ten with *schioppi*, and three or four thousand Indians, our friends'.[53] Even after new supplies arrived, proportions remained similar, so that another force had 'three hundred infantrymen and fifty between crossbowmen and *schioppettieri*'.[54] As Jacob and Visoni-Alonzo have observed, in the Spanish conquest of the Americas, beyond the 'initial psychological factor', firearms had in general 'a very limited effect'.[55]

That did not deter travellers to and observers of the Americas from emphasizing the superiority of the firepower they did have. Allegretto Allegretti of Siena described Indigenous people who did not have 'any type of weapons'; a Venetian writer commented that they used 'the tips of reeds on top of sticks' due to the lack of available iron.[56] The motif is repeated in the fake Vespucci letters, which recorded that 'they have no arms but bows and darts'.[57] The anonymous narrator of Pedro Alvares's 1500 voyage who reported the people of Calicut marvelling greatly at their gun salute had also stopped on what would turn out to be the coast of Brazil. 'In this land,' he wrote, 'we did not see iron, nor other metal; they cut wood with stone.'[58] Giovanni da Empoli confirmed that 'their arms are like darts, the tips covered with fishbones'.[59] The people Alvares's crew had encountered were the Tupinambá, who did indeed live without iron tools. Pietro Martire confirmed that the people Columbus and his crew encountered further north, in the Caribbean, likewise 'lacked iron'.[60] By the time the *On the New World* was published, it could include such descriptions as the Indigenous people being 'armed in their usual style', a usage in the text that assumes readers' familiarity with the limitations of Indigenous weapons.[61] European readers of the 1530s were also evidently expected to be aware of typical weapon ranges (a *schioppo* shot and a half; two crossbow shots; one bow shot), which Oviedo used in his *Natural and General History* as a means of measurement.[62]

While some Indigenous armour protected against crossbow bolts, guns were a different matter.[63] Martire recorded several incidents in which *schioppi* contributed to Spanish conquests: on one occasion the Spanish, 'having discharged some *schioppi*, made [the Indigenous people] flee and turn their backs and abandon that place where they lived'; on another occasion, hearing the noise of firearms, 'they thought that they were lightning bolts from the sky, and they were set to such flight and fear that many of them fell to the ground'.[64] Even when Indigenous people did not apparently experience fear, the outcome was still a European victory: 'neither arrows nor artillery fire from the ships . . . could frighten them, and they judged it better to die than see their fatherland occupied'.[65] In the *Natural and General History*, Oviedo described an incident in which 'one of our *schioppettieri* sent an Indian to the ground with a shot'. This seemed to be an important man, he wrote, 'because the Indians quickly lost heart and moved their army a little way behind where the *schioppetto* couldn't reach'.[66] While on the one hand this observation illustrates that Indigenous troops had become familiar with the range of guns, on the other it underlined that the killing of individual commanders, whether deliberate or incidental, could have an impact on enemy morale. The guns that in Europe were frowned upon for enabling 'cowards and shirkers' to kill at a distance here acquired a more positive gloss.

It was arguably during these earliest encounters that handguns afforded the greatest advantage. A 1543 letter from Oviedo to Cardinal Pietro Bembo described how

> in some encounters and battles, of which there were many, certain Spaniards were killed, and they [the Spaniards] killed many more of the Indians, because the less familiar they [the Indians] were with arquebuses and crossbows, the more recklessly they were killed by those arms: and some thought that the bangs and noise and stink from the arquebus were bolts from the sky, and seeing the damage in many places they immediately ran away; in many others they waited and opposed with great ardour in their own defence and that of their land.[67]

This theme finds an echo in Michel de Montaigne's 1580 description of firearms instilling terror: 'the lightning flashes of our cannons, the thundering of harquebuses [would have been] able to confuse the mind of Caesar himself in his day if they had surprised him when he was as ignorant of them as [the peoples of the Americas] were'.[68] By the time of Montaigne, this was firmly established as a trope of European accounts of conquest. It appears on multiple occasions in Ramusio's volumes.[69] It no doubt had an element of truth: as we saw in chapter 5, the use of firearms to intimidate became an important feature of continental European gun culture away from the battlefield. Indeed, by repeatedly underlining the potential of firearms to threaten enemies, these texts reinforced and perhaps even contributed to the prevailing repertoire of European ideas about guns and their functions. It is worth reiterating here that these European accounts, of course, typically write out the Indigenous perspective. David J. Silverman has argued (in relation to the early seventeenth century) that Native Americans were not so much intimidated by firearms as intrigued 'about what they could accomplish with European weaponry'. It took a while for a market to grow up, but in that case it was Iroquois gunmen, not the colonizers, who 'galvanized an arms race throughout the Native Northeast'.[70]

A limiting factor in handgun use was the ability to produce sufficient gunpowder. Although crossbow bolts could be manufactured, according to the account of Francisco de Ulloa (1540 or earlier), powder supply remained challenging until local resources for gunpowder production were identified.[71] Indeed, a recurring theme in the narratives collected by Ramusio is the need for effective maintenance and supply of arms. Oviedo noted in 'a certain valley on the island of Cuba' the presence of a mine producing stone suitable for use in bombards and *schioppetti*.[72] The island of Cubagua (located off the coast of Venezuela) was 'all saltpetreish' (*tutta salnitrosa*).[73] In a letter of 1524 from Guatemala, Pedro de Alvarado reported to Hernán Cortés that

> in this village there is a mountain of alum, a mountain of vitriol and another of sulphur, the best that has so far been found, and from which, with a piece that was brought to me, without refining it or doing anything else to it, I extracted seventeen pounds of very good powder.[74]

An anonymous Spanish relation, sometimes attributed to a companion of Cortés, further described the presence in the area around Tenochtitlan of 'mines of gold and silver, copper and tin, steel and iron', an assessment that was more enthusiastic than accurate but illustrates the importance attached to securing metals.[75]

Oviedo's text is also notable for his use of religious metaphor. For example, in a passage on the riches of the Indies, he observed that 'those people who possess so much gold do not have poisoned darts, nor do they know of *schioppi*, powder, instruments of war and arms, whether defensive or offensive, and so they flee from a horse like demons from the cross'.[76] This description shifts the usual diabolical metaphor of firearms: here European arms are neutral to positive, while the people they are used to target become the demons. Even though the colonists often operated with very few firearms (in one battle, says Oviedo, only two *schioppettieri* were present), on one occasion when they were trying to secure a water supply, gunpowder weapons saved Christians from their deaths:

> And if it wasn't for the artillery and for those few crossbowmen and *schioppettieri* that were among us, more Christians would have

> perished, because they could not use other weapons than the ones already mentioned. And it is believed that those artillery shots and the crossbowmen did a lot of damage and killed many Indians, but the number cannot be known, although some were seen falling and therefore there was great fear in them.[77]

Similar rhetoric appears in an account of Portuguese diplomatic discussions of the 1520s with the representatives of Dawit II of Ethiopia, in which the participants considered whether the Portuguese or the 'Moors' were more frightened by artillery fire. Alvarez, the Portuguese representative, assured his hosts that 'we were not scared at all, because we fought with the faith of Jesus Christ'.[78] Moreover, when in 1524 the troops of Cortés succeeded in identifying supplies of metal and saltpetre, he wrote: 'God has provided for munitions, we having found so much and such good saltpetre that we will be able to make provision for other necessities.'[79] This was a far cry from the accounts of the European wars being written at that time, which were still commonly placing guns and gunpowder on the side of the Devil.

From Mexico to Florence

The description of Cortés's palazzo in Mexico City published by Oviedo in his *Natural and General History of the Indies* both illustrates how these natural resources were subsequently exploited and sheds light on an important European representation of gunpowder production. Inside the palazzo, 'there were rooms for ammunition and artillery, and rooms with great quantities of arms and armour, and there were stables for two hundred horses, and workshops for making artillery powder, and seven or eight smithies which continually made new weapons and crossbows'.[80]

This description provides important context for the frescoes commissioned from Ludovico Buti in 1588 for the Uffizi Armoury (Figures 9.3 and 9.4 show small sections of the whole).[81] The sequence of ceiling paintings for the different rooms shows the production of armaments and the conquest of Mexico, as well as a range of other Florentine and European battles. Well-known to scholars, the

FIGURE 9.3 Ludovico Buti, *Manufacture of Gunpowder*, 1588, Florence. Fresco, Gallerie degli Uffizi, Florence. © NPL – DeA Picture Library / Bridgeman Images.

representation of Amerindians in this series has been analysed in some detail, rather briefly by Detlaf Heikamp half a century ago and more recently in greater depth by Lia Markey; between them they have identified a number of sources for it, importantly the *Florentine Codex*.[82] Before turning to the frescoes themselves, it is worth considering how guns were represented in the *Codex*.

The *Codex* was an important source for cultural understandings of firearms, at least at the Florentine court. A manuscript compiled by Fray Bernardino de Sahagún and his collaborators (including the Nahua scholars Antonio Valeriano, Martín Jacobita, Pedro de San Buenaventura and Alonso Vegerano) between the late 1540s and 1577, it was acquired by Cardinal Ferdinando I de' Medici sometime before his elevation to grand duke in 1587. It drew on first-hand testimony from Indigenous veterans (albeit at some decades from events) and provided important evidence of their experience of Spanish conquest and therefore of those faced

FIGURE 9.4 Ludovico Buti, *Spanish Troops Conquering Mexico*, 1588, Florence. Fresco. Gallerie degli Uffizi, Florence. © NPL – DeA Picture Library / Bridgeman Images.

with guns for the first time.[83] The cardinal commissioned an Italian translation that would have facilitated a wider diffusion of the text, although the context of pressure to destroy 'idolatrous' writings on Indigenous religion restricted its circulation to a trusted circle.[84] The text is complex, consisting as it does of two alphabetic texts (Nahuatl and Spanish) plus a series of images, each conveying a somewhat different story.[85] While the images were drawn by Nahua artists, they were influenced by western art, very probably by the images of Charles V and his troops at the Battle of Mühlberg discussed in earlier chapters.[86] An inconsistency is apparent, for example, in the text and illustration to Chapter 5 of Book 12, which describes an encounter between the messengers of the Aztec emperor Moctezuma and Hernán Cortés, on the latter's boat. Cortés's men restrained the messengers with iron manacles, then shot a 'great lombard gun' or a 'cannon' (the translators give this somewhat differently), but the accompanying image shows a handgun

(Figure 9.5a). The messengers 'fainted away and swooned' in response and were subsequently revived by the Spanish with wine and food.[87] The intent to intimidate is clear; that it worked is clear from the subsequent account of Moctezuma's response when the messengers reported back:

> And when he had heard what the messengers reported, he was terrified, he was astounded. . . . Especially did it cause him to faint away when he heard how the gun, at [the Spaniards'] command, discharged [the shot]; how it resounded as if it thundered when it went off. It indeed bereft one of strength; it shut off one's ears. And when it discharged, something like a round pebble came forth from within. Fire went showering forth; sparks went blazing forth. And its smoke smelled very foul; it had a fetid odor which verily wounded the head. And when [the shot] struck a mountain, it was as if it were destroyed, dissolved. And a tree was pulverized; it was as if it vanished; it was as if someone blew it away.[88]

Again, the description here is somewhat inconsistent so far as firearms are concerned: the 'pebble' suggests a relatively small weapon, but it seems unlikely that an arquebus could destroy a mountain or pulverize a tree. The text as a whole features references to both large and small guns (including the discussion of arquebuses in the conquest of Tenochtitlan cited in the introduction): we hear of Aztec fighters trying to dodge Spanish shooters; the Spanish shooting their arquebuses while climbing the city's pyramid temple; the looting of war goods; and shooting from boats.[89] The penultimate chapter, recounting the submission to the Spanish of Moctezuma's nephew and heir Quauhtemoc, likewise features guns: first a gun salute from a roof terrace over the heads of the 'common folk', and then shots presumably into the crowd, during which 'many died'. 'Just so,' the *Codex* records, 'did the war go to its end.'[90]

The images of the *Codex* also provide vivid evidence for the presence of firearms in the conflict. The opening image of Book 12 shows in the foreground, on the ground, a gun and powder-flask (Figure 9.5b). This is imagery that echoes artworks such as the Avalos tapestries, but it is notable here that while—as Kevin Terraciano has argued—the image draws on imagery of the loading of

FIGURE 9.5A–E *General History of the Things of New Spain by Fray Bernardino de Sahagún: The Florentine Codex. Book XII: The Conquest of Mexico*, 1577, Mexico. Biblioteca Medicea Laurenziana, Florence. Library of Congress World Digital Library image numbers 21, 1, 26, 100, 75 (details). Public domain.

ships for the Trojan War, the artist has made a positive choice to underline the presence of this technology.[91] Elsewhere firearms are similarly emphasized. The image of the Spanish in war array features an Aztec man pointing directly to the arquebus, highlighting its presence in a way that the pikes and cannon also featured are not

FIGURE 9.5B (*continued*)

(Figure 9.5c). Images of the Spanish marching on Tenochtitlan feature cavalry with crossbows and infantry with handguns, as would be expected, but several also illustrate cavalrymen with handguns (Figure 9.5d), an unusual use prior to the development of reliable wheellocks which might be interpreted as either improvisation or development of novel military technique.[92] As the narrative moves to the outbreak of war between the Spanish and Aztecs, a Spaniard in helmet and body armour is shown shooting down from behind a wall at an Aztec, smoke emerging from the weapon (Figure 9.5e). This may be a simple representation of a routine military tactic, but I am struck by the similarity to the imagery of demons shooting down at Christ that we saw in chapter 1 (Figure 1.8). Just as Cortés's

FIGURE 9.5C *(continued)*

thanks to God for gunpowder flipped the usual European narrative of guns, here the Spanish occupy the position once accorded to the demons.

Further images in the *Florentine Codex* include a gun being shot, and a gun lying on the ground as Aztecs loot other items abandoned by the retreating Spanish, though notably not firearms.[93] Finally, multiple images show the presence of guns on ship, sometimes alongside larger artillery; sometimes fired from shipboard at targets on land; sometimes used immediately on landing.[94] Overall, the images do illustrate the mixed military technology deployed in the course of conquest.[95] However, they give a prominence to handguns that arguably confirms their importance in collective memories of the conflict, an importance that came to be reflected in European representations too.

FIGURE 9.5D (*continued*)

Turning to the Uffizi frescoes, echoes of the *Codex*'s representation of guns are clearly apparent in Buti's images of battle, notably the unusual iconography of a gunman on horseback (Figure 9.4). Moreover, in light of guns' prominence in the *Codex* and in Oviedo's description of Cortés's palazzo, the depiction of armament production in the first room of the Armoury should be considered in relation to the images of the conquest elsewhere in the

FIGURE 9.5E *(continued)*

sequence. Assessing representations of Mexico, the prior studies focused on the latter. Yet, taken together, the two ceilings tell a story, first of preparation for war—the images feature the production of armour, swords, cannon and gunpowder as well as designs for fortifications—and then of the deployment of weapons in conflict. (Handgun production is not shown in this design, but then it is not specifically mentioned by Oviedo, and, as we have seen, gun barrels required highly specialized production.) The ceilings are linked by design features: the shape of the composition as a whole is consistent between the two rooms, the frescoes in one showing arms production and in another the conquest, as are elements of the borders. By the time of the commission, the relevant section of Oviedo's *Natural and General History* was certainly available to readers in Florence via the Ramusio edition, and while the frescoes do not hint at a specific location for the production spaces they depict, the sequence is certainly open to both the particular reading that these are the rooms used to produce the weapons used in the successive events, and to a more general reading that European arms production enabled the conquests. It is all the more plausible that a link was intended given the established practice in the Medici redevelopment of the Palazzo Vecchio of taking seriously the representation of arms and armour. As we saw in Chapter 8, Giorgio Vasari's frescoes for the Sala del Cinquecento tell a story of the development of weaponry from the Battle of Pisa to the Battle of Siena. Related interests are also apparent in the design for the *studiolo* of Grand Duke Francesco I (r. 1574–87), which features images of mining and the forge. In decorating an Armoury intended for the display of arms and armour to diplomatic visitors among others, careful thought was surely given to the scheme as a whole, given the typical intention for such works of prompting conversation.

Conclusion

The story of guns in European imperial expansion is a complex one. While the reality was often of swift local adoption of firearms, representations back on the continent often told—subtly or overtly—a tale of technological superiority. Importantly, they did so at an early stage of guns' wide adoption, contributing to a more positive image for firearms in a society that often remained ambivalent. Whatever their origins, guns and gunpowder were narrated as European inventions that characterized the continent's modern times. For all the scholarly challenge to the idea that guns really were decisive, the idea of technological superiority facilitating conquest retains a grip on the popular imagination to this day: its very longevity may provide an explanation for its persistence. In fact, however, it is more accurate to say that conquest enabled guns than that guns enabled conquest.

Yet if this was a story Europeans told themselves, it was also a story told *about* them, not least in some of the earliest

FIGURE 9.6 Unknown Edo maker, figure of Portuguese soldier, late 16th to early 17th century, Benin City. Lost-wax cast in brass. 37.5 cm height. British Museum, Af1928,0112.1. © The Trustees of the British Museum.

FIGURE 9.7 Master of Calamarca, *Asiel, Fear of God* (*Asiel, Timor Dei*), before 1728. Oil on canvas and gilding, 160 × 110 cm. Museo Nacional de Arte, La Paz, Bolivia. Photo: PicturesNow / UIG via Alamy.

three-dimensional sculptures of Europeans with guns. The Benin bronze shown in Figure 9.6, made by Edo sculptors in what is now Nigeria, was produced in the late sixteenth or early seventeenth century and shows a Portuguese soldier bearing a gun; other bronzes show crossbows, illustrating again the mixed technologies present.[96] The bronzes have recently been the focus of much attention over their looting by British troops in 1897 and requests for their return.[97] However, they should also be recognized for what they show—which is that before any European sculptor put a gun in the hand of a subject, and while the continent's painters remained cautious about depicting firearms in individual portraiture, the artists of the Benin court had made the association. The earliest Portuguese contact with the kingdom was probably around 1486, when some trading links were established, and the figures of European soldiers have often been associated with the rule of Oba Esigie (r. c. 1504–50), whose alliance with the Portuguese facilitated his rise to power.[98] Even more so than the evidence from the Americas, however, they indicate the significance accorded to these weapons by an non-European power: the difference is perhaps a product of the relative importance of guns in West Africa, where they were well suited to regional patterns of infantry warfare. Like the evidence from veterans that informed the Florentine Codex, they offer a vital outside perspective on Europeans and their guns. So, from the late seventeenth century, would the iconography of armed archangels (Figure 9.7). Found in what are now the borderlands of Bolivia and Peru, these paintings brought together the Christianity of European conquest with its most iconic weapon: the gun.[99] No longer was this the weapon of demons, but of angels.

LATE 1. Bernard van Orley (design), Willem and Jan Dermoyen (tapestry), *The Battle of avia: Advance of the Imperial Army*, c. 1525–31. Capodimonte Museum, Naples. In the ackground, arquebusiers lower their guns to fire.

PLATE 2. Bernard van Orley (design), Willem and Jan Dermoyen (tapestry), *The Battle of Pavia: Defeat of the French Cavalry* (detail), c. 1525–31. Capodimonte Museum, Naples. The Spanish army scored notable victories in the early 1500s through its effective use of handguns.

PLATE 3. Unknown artist, *The Battle of Anghiari* (detail), late 1460s. National Gallery of Ireland, Dublin. This battle scene decorated a storage chest, bringing images of guns into the domestic sphere.

PLATE 4. Nuno Gonçalves (design), attr. Pasquier Grenier (manufacture), *The Siege of Asilah* (detail), Tournai, 1471. Colegiada de Pastrana, Guadalajara. This tapestry shows both the Portuguese victors and their Moroccan opponents using handguns.

PLATE 5. Robinet Testard, *Hours of Charles of Angoulême*, 1480–96, fol. 110v (detail). Bibliothèque Nationale de France. This image of demons shooting at the risen Christ was included in a French religious text.

PLATE 6. Unknown maker, powder-flask with bullet box, clock, compass, and sundial, Augsburg or Nuremburg, c. 1570–1600. Metropolitan Museum of Art, New York. The decoration shows the conquest of New Carthage by the Roman general Scipio.

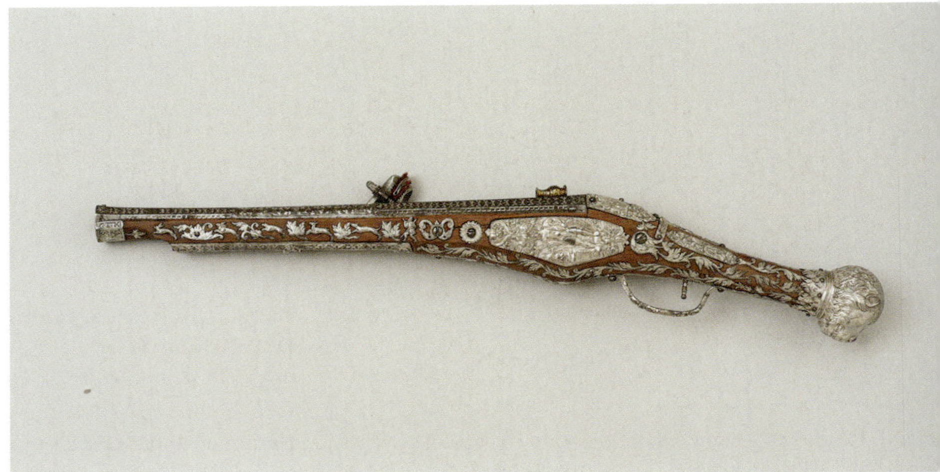

PLATE 7. Unknown maker, wheellock pistol for Archduke Ferdinand II of Tyrol, Brunswic c. 1555. KHM, Vienna. The decoration combines foliage designs and hunting scenes with both Christian and classical imagery.

PLATE 8. Peter Peck (gunsmith), Ambrosius Gemlich (etcher), double-barrelled wheelloc pistol made for Emperor Charles V, Munich, c. 1540–45. Metropolitan Museum of Art, New York. One of the earliest surviving wheellocks, this luxury weapon incorporates Charles's insignia.

ATE 9. Unknown maker, wheellock breech-loading pistol, ?Augsburg, c. 1545. KHM, enna. The subtle decorative style here follows the wider aesthetics of the Renaissance mestic interior.

PLATE 10. Unknown maker, combination axe-pistol of Grand Duke Ferdinand I de' Medici, Germany, c. 1580. Metropolitan Museum of Art, New York. Novelty weapons like this were popular at the Renaissance courts.

PLATE 11. Jan van der Straet / Giovanni Stradano, *St John's Day Fireworks in Piazza della Signoria* (detail), c. 1558. Palazzo Vecchio, Florence. Both orderly and apparently disorderly shooting are shown in this image.

PLATE 12. Jan Cornelisz Vermeyen (design), Willem de Pannemaker (tapestry), *The Conquest of Tunis* (detail), 1548–54. Royal Palace, Madrid. A gun is prominent in the foreground of this image of Imperial soldiers gambling for booty.

PLATE 13. Unknown artist, *Kenau Simonsdochter Hasselaer*, c. 1590–1609. Rijksmuseum, Amsterdam. Hasselaer was mythologized for her defence of Haarlem during the 1573 siege. A wheellock gun is shown among her weapons.

PLATE 14. Unknown artist, *St Charles Borromeo, Having Survived an Attack*, 17th century. Chiesa di S. Michele e della Madonna di Biorca, Sondrio. This is one of several artworks depicting the 1569 shooting of Carlo Borromeo, archbishop of Milan.

PLATE 15. Franz Hogenberg, *Death of Admiral Gaspard II de Coligny*, c. 1572. Bibliothèque Nationale de France. Coligny survived the shooting shown on the left, only to be murdered by stabbing less than two days later.

PLATE 16. ?Jacques Daret, *Histories of Alexander the Great*, Tournai, c. 1460. Villa del Principe, Genoa. This tapestry includes a rare depiction of a Black handgunner in European art.

PLATE 17. ?Jacques Daret, *Histories of Alexander the Great* (detail), Tournai, c. 1460. Villa del Principe, Genoa. The gunfire is clearly shown in this detail.

PLATE 18. Unknown artist, after Titian (Tiziano Vecellio), *Alfonso d'Este, Duke of Ferrara*, late 16th or early 17th century. Metropolitan Museum of Art, New York. This painting alludes to Alfonso's early interest in gunpowder technology.

PLATE 19. Titian (Tiziano Vecellio), *Emperor Charles V at Mühlberg*, Venice, 1548. Museo del Prado, Madrid. Charles wears a wheellock pistol at his hip, in the style of his light cavalry.

PLATE 20. Giuseppe Arcimboldo, *Fire*, 1566. KHM, Vienna. This allegorical image alludes to the Emperor Maximilian II and incorporates cannon, a handgun barrel, a powder-flask and match-cord.

PLATE 21. Unknown maker, majolica dish showing Battle of Mühlberg, Urbino, c. 1555–60. Victoria and Albert Museum, London. One of many images of this battle, produced across multiple media, which emphasize the use of firearms.

PLATE 22. Pieter Bruegel the Elder, *The Conversion of Paul* (detail), Netherlands, 1567. KHM, Vienna. Most European art showing biblical scenes does not feature guns, even when the figures are depicted in modern dress. This work is an exception.

PLATE 23. Jan Cornelisz Vermeyen, *Tapestry Cartoons for the Conquest of Tunis by Charles V: An Unsuccessful Sortie by the Turks from La Goletta* (detail), 1546–50. KHM, Vienna. The Ottoman cavalry here are shown with firearms.

PLATE 24. Giorgio Vasari, *The Battle of Marciano*, 1567–71. Palazzo Vecchio, Florence. This fresco series conveys the growing importance of firearms in 16th-century military campaigns.

PLATE 25. Unknown artist, *Portrait of Robert Dudley*, 16th century. Parham House, West Sussex. Dudley poses with a petronel in a portrait alluding to his role leading English troops in the Netherlands.

PLATE 26. Attr. Steven van der Meulen, *Portrait of Thomas Butler, 10th Earl of Ormonde*, mid-16th century. National Gallery of Ireland, Dublin. Shown here with a wheellock gun, Butler was a key figure in the Tudor conquests that paved the way for the colonization of Ireland.

PLATE 27. Cornelis Ketel, *Portrait of Sir Martin Frobisher*, 1577. Bodleian Libraries, Oxford. Frobisher is best known for his three voyages made in search of the North-West Passage. This portrait was commissioned by the Cathay Company.

PLATE 28. Bartholomeus van der Helst, *Militia Company of District VIII in Amsterdam under the Command of Captain Roelof Bicker* (detail), Amsterdam, c. 1640–43. Rijksmuseum, Amsterdam. The presence of the single unarmed Black boy in this painting points to the wider imperial context for gun use in the 17th century.

PLATE 29. Unknown maker, *Continuation of the Triumphal Procession: Presentation of War Material*, Brussels, 1555–60. KHM, Vienna. This work marked Dom João de Castro's 1547 conquest of Goa and contributed to a wider European narrative that emphasized the importance of military technology in imperial expansion.

PLATE 30. Ludovico Buti, *Manufacture of Gunpowder*, 1588. Uffizi Gallery, Florence. Part of the ceiling of the first room in the Uffizi Armoury, which was dedicated to images of armament production.

PLATE 31. Ludovico Buti, *Spanish Troops Conquering Mexico*, 1588. Uffizi Gallery, Florence. The subsequent rooms in the Uffizi Armoury showed firearms in action, here in the context of European imperial expansion.

PLATE 32. Master of Calamarca, *Asiel, Fear of God*, before 1728. Museo Nacional de Arte, La Paz. Paintings of armed archangels like this one brought together the Christianity of European conquest with its most iconic weapon, the gun.

Conclusion

A CASUAL OBSERVER of early modern European culture would be forgiven for missing the presence of guns. None of the many deaths in the plays of Shakespeare are caused by gunshot. Far more attention has been devoted to the impact of print than to firearms, although many people who never read a book or pamphlet must have fired a gun. Yet over the course of the sixteenth century, the handgun made the transition from novel and decisive military technology to become an everyday object, in use across society and carrying a new set of cultural associations that would persist through the coming centuries. This was the firearm revolution. The history of guns offers an important window on the priorities and internal tensions of the early modern state. Rather than separating out the state's military function from its role in enforcing social discipline, and from the idea of a civilizing process, a study of the gun brings the three together, offering a holistic approach to early modern state formation. If you imagine a Venn diagram of these three theories, in the space where they overlap we find—and the people of early modern Europe found—a gun.

Attitudes towards guns were initially characterized by hostility, and to some extent this persisted even through their widespread adoption. In the literary sources at least, firearms remained a coward's weapon and the Devil's work. Rather few users wrote enthusiastically about guns, although some did, and of course the written sources represent the experiences of a limited social stratum. There were, however, certain contexts in which firearms might be perceived more positively, notably in quashing religious rebellion. Indeed, the imagery both *on* guns and *of* guns suggests the religious wars of the 1540s were a key turning point in making the new weapon godly. Religious justifications were also important to imperial expansion, and while in reality the role of firearms in conquest was limited at best, the stories of guns' shock value bolstered beliefs within Europe about the continent's superiority, technological and otherwise. They also contributed a new and more positive narrative of firearms at a crucial point in Europe's adoption of this contentious technology. By mid-century handguns were thoroughly assimilated into European culture.

Firearms demonstrate the importance of attending to the relations between states and peoples: a state that trained people to shoot, required them to engage in militia drill, and expected them to prepare to fight in defence of the fatherland could not then simply require an about-turn on ownership of those weapons perceived as most effective. The sixty-five years of on-off conflict in Italy from 1494 to 1559 with their associated demands for militia service, plus wider opportunities for soldiering, created an environment that was well adapted to the subsequent proliferation of guns. Competition between the European and Mediterranean powers, highly apparent on the divided Italian peninsula, was a key factor in promoting this new and lethal technology and overrode anxieties about social order. Citizens learned how to use firearms; indeed, they were positively incentivized to do so. The widespread use of private contracting in wartime along with chaotic demobilization processes made it easy for guns to spread. Once the population became accustomed to using firearms, it was not a great step for them to prefer the wheellock as a weapon better adapted to self-defence. Elite resistance to gun laws illustrates the challenges confronted by states legislating for social discipline in the face of older forms of power derived both from feudal landownership arrangements and from relations of patronage in court or city contexts. Once exemptions were conceded, pressure grew to extend them to men outside the nobility, who now perceived themselves to be under threat. Governing regimes were well aware of the techniques that might be used to check proliferation in theory, from bans on the sale and gambling away of weapons to licensing arrangements and police activity to seize illegal weapons. The more radical solutions aimed at gun producers were, however, tried only briefly in the Papal States. The anonymous author of a proposition for international gun control quoted in the introduction was correct to suggest that it would take a multilateral agreement to make them stick. In the absence of any such solution, however, the pressure to maintain production to facilitate defence—along with a population trained in arms use, fearful of attack, and desirous of access to weapons for personal protection—militated in one direction only: that of further gun proliferation. Once set in motion, the firearm revolution acquired its own momentum. Gun control in Europe was only effectively implemented centuries later, by much larger and more interventionist states.

Paradoxically, however, guns may have driven a civilizing process, even if that process was fragile. Widespread availability of a lethal technology required control and restraint on the part of its users, although in practice that was more apparent in some environments than others. Any breakdown in social order would now have a new and more deadly dimension. However, by prioritizing effective defence and distributing weapons during the Italian Wars, states contributed to a process that undoubtedly contributed to the violence they struggled to quash in the second half of the sixteenth century. Moreover, as the quality of firearms improved, that opened the way to a new type of political violence. Assassination no longer necessitated close proximity to the target: a prominent person could reliably be shot from a window.

Within the world of warfare, guns also made for change. This was not only true on the battlefield, where new tactics were developed to maximize their effectiveness, but in terms of manufacture. Guns required specialist production, which in the case-study discussed here was contracted out to private producers, for whose services there was evidently competition. So far as the state was concerned, the complex contracting processes drove a need for oversight (whether or not it was always successful). The limited source material means it is hard to map the detailed dynamics of the small arms industry's development at this early stage, but there is certainly scope for further research on its economic history. Finally, the early history of guns poses important questions about how societies should respond to new technologies. It is very clear that in this case the technology ran decades ahead of any careful debate within government about how it should be managed. Enthusiasm for the military technology trumped any concerns that might have been raised about its broader implications, even as plenty of prominent voices lamented the rise of the 'wretched' weapon.

The ideas about guns that took shape as the firearm revolution played out framed subsequent debate about small arms and their use both within Europe and beyond. This is not to say that sixteenth-century attitudes were precisely the same across the continent: some places had more open gun cultures than others; militia and guilds did not work in identical ways. Still, it is possible to draw out a set of themes formed in these crucial decades for gun proliferation that continued to shape thinking on guns over centuries. These include their importance as a military technology, associations with ideas of liberty and defence of the fatherland or republic, an idea that they can be useful in self-defence, a negative association with the Devil or the underworld that was still in use by critics of the arms industry in the twentieth century,[1] and an understanding that they had played a key role in European conquests around the world. This cultural legacy of the firearm revolution has been remarkably persistent.

Today, there are over a billion guns in the world, 85 per cent of them in private hands.[2] Beliefs about guns remain contested. Some people are deeply attached to their weapons and others repelled by the idea of shooting, but attitudes towards guns can change over time, not least with shifts in assessment of how far we can trust the state. Different states opt for different levels of gun control: even within the continent of Europe today there are different norms of policing and distinctive gun cultures. Debates about to whom arms should be exported, how many and for what uses, continue within and outside governments. If we no longer need luxury weapons to help assimilate a new technology in elite environments, we still have sporting contests—at the Olympics, for example—that offer a platform for gun manufacturers to display their brands. The film industry fulfils a role in telling stories about guns that once were told in pamphlets and poetry. I hope that by going back to the original processes through which society adopted guns, we might find new ways of approaching the arguments that persist today. If I have made you stop and think again about any aspect of the gun—or indeed about any risky technology—then this book will have succeeded.

ACKNOWLEDGMENTS

I FIRST WROTE about the history of firearms when I was a master's student at Royal Holloway, University of London, in a course taught by Hugo Blake and Sandra Cavallo. That was in 2002. Doctoral research took me elsewhere, but I returned to the topic of guns in the mid-2010s while writing a biography of Alessandro de' Medici: there was still limited scholarly literature with which to contextualize Alessandro's gun collection, and eventually I decided I had better write something myself. I am grateful to the Arms and Armour Heritage Trust for funding the initial archive research for this book, and to Swansea University and Manchester Metropolitan University for further research support, including sabbatical leave. As part of the Manchester Met RISE undergraduate development programme, Ophir Barak assisted with bibliographical research in the summer of 2021. A Balsdon Fellowship at the British School at Rome in the autumn term of 2023 enabled me to finalize research in the Italian archives and view important visual sources in Naples and Florence for myself. I would like to thank my agent, Catherine Clarke, for arranging publication with Princeton, and Ben Tate and his colleagues for their support through the publication process.

Preliminary drafts from this book were presented to the Late Medieval and Early Modern Italy seminar at the Institute for Historical Research; the Society for the History of War 'New Voices' conference at All Souls College, Oxford; and at Aberystwyth University, the University of Edinburgh and Queen Mary University of London. Further papers were presented at the 2022 Society for the History of War conference in Amsterdam, the 'Worlds of Conflict' conference at the University of Toronto in 2023, and for the Princeton University early modern graduate seminar and the University of Pittsburgh 'Gun Violence and Its Histories' initiative in 2024. I thank the many colleagues who commented on those occasions as well as the peer reviewers who commented on both the manuscript itself and on the previously published sections of the book. I'm sorry I don't have space to name you all individually. Chapters 2 and 3 incorporate material first published in 'Agents of Firearm Supply in Sixteenth-Century Italy: Constructing the Military State' in the collection *Shadow Agents of Renaissance War* edited by Stephen Bowd, Sarah Cockram and John Gagné. Chapters 5 and 6 incorporate material previously published in 'Firearms and the State in Sixteenth-century Italy: Gun Proliferation and Gun Control' in the journal *Past & Present* (2023), and forthcoming in the conference proceedings *Cities in Arms: Urban Martial Cultures in Late Medieval and Early Modern Europe* edited by Regula Schmid and Daniel Jaquet.

I am particularly grateful to Peter Wilson and his team at the Fiscal-Military Hubs project for discussion over the past few years, as well as to Mark Bennett at the Royal Armouries, and everyone involved in the Arms in the Aftermath of Conflict initiative, not least Rosie Johnston for her hospitality on my trips to Prague and Vienna. The doctoral research of Joe Tryner and Victoria Bartels (as well as conversations with them) proved enormously helpful to this project. Raphael Beuing and Marius Mutz kindly arranged a private visit for me to the Bayerisches Nationalmuseum in Munich, and Jana Sedláčková did likewise at Konopiště Castle outside Prague. Dario Zago showed me round the Museo Luigi Marzoli in Brescia, while Amanda Madden kept me company in Venice and very helpfully commented on an early draft of the book. Jonathan Ferguson at the Royal Armouries deserves a special mention for allowing me to try shooting a replica sixteenth-century matchlock, as do the instructors and participants at the Shotgun and Chelsea Bun Club for fun mornings out trying to hit clay pigeons. Finally, thanks to my old friend Liz Yeates for keeping me company at shooting school, even if you did prove a much better shot than me.

NOTES

Abbreviations Used

ARCHIVAL SOURCES

ASBo	Archivio di Stato di Bologna
ASBs	Archivio di Stato di Brescia
ASC	Archivio Storico Capitolino (Rome)
ASCBs	Archivio Storico Civico di Brescia
ASE	Archivio Segreto Estense (in ASMo)
ASF	Archivio di Stato di Firenze
ASMn	Archivio di Stato di Mantova
ASMo	Archivio di Stato di Modena
ASNa	Archivio di Stato di Napoli
ASP	Archivio di Stato di Parma
ASR	Archivio di Stato di Roma
ASS	Archivio di Stato di Siena
ASVe	Archivio di Stato di Venezia
CCX	Capi del Consiglio dei Dieci (in ASVe)
CPI	Cancelleria prefettizia inferiore (in ASBs)
GM	Guardaroba Medicea (in ASF)
KHM	Kunsthistorisches Museum
MAP doc.	Medici Archive Project Document (in ASF)
MdP	Mediceo del Principato (in ASF)
SG	Soldatesche e Galere (in ASR)

ITALIAN ARCHIVAL ELEMENTS

b.	busta (folder)
bb.	buste (folders)
cat.	catalogo (catalogue)
fasc.	fascicolo (file)
ins.	inserto (insert)
reg.	registro (register)
regg.	registri (registers)

Introduction

1. Sabatti, 'La Valtrompia nel '500', 165, includes an image but unfortunately not an archival reference. The date was not mentioned by Morin and Held in their 1980 history of the company (*Beretta*) but has been in public circulation since at least 1982; see Tagliabue, 'Small Steel Mills Feel Pinch in Italy'. For more recent use, see R. L. Wilson, *World of Beretta*, who says it is in the Record Office Archives, Venice, and describes it as a 'document of account from the Senate of the Republic of Venice', but again without more specific citation.

2. Giovio, *Notable Men and Women*, 57.

3. Cardano, *Book of My Life*, 166.

4. Grafton, Shelford and Siraisi, *New Worlds, Ancient Texts*, 203; Chase, *Firearms*, 197.

5. Satia, *Empire of Guns*.

6. Hale, 'Gunpowder'; Brugh, *Gunpowder*; Bartels, 'Masculinity, Arms and Armour'.

7. Elias, *Civilizing Process*.

8. Rose, *Renaissance of Violence*; Carroll, *Enmity*.

9. Archivio di Stato di Bologna (hereafter ASBo), Legato, Licenze di portare armi, fol. 47v.

10. The proposition, a document of three folios and about two thousand words, is in Archivio di Stato di Firenze (hereafter ASF), Miscellanea Medicea 39, ins. 15; another copy is included in a seventeenth-century collection of discourses on military matters now in the Biblioteca Augusta of Perugia.

11. Sahagún, *Florentine Codex*, 40.

12. Bowd, *Renaissance Mass Murder*; Davis, 'Up in Smoke'.

13. Carr, *Kalashnikov Cultures*; Nowak and Gsell, *Handmade and Deadly*.

14 Eisenstein, *Printing Press*, ch. 1.

15. For the guns discussed here, the most important reference work of this type is Morin, *Armi antiche*.

16. Schwoerer, *Gun Culture*, 5, 106; Tlusty, *Martial Ethic in Early Modern Germany*; Delle Luche, 'Pouvoir, fabricants et sociétés de tir'.

17. Brugh, *Gunpowder*; Nayar, *Renaissance Responses to Technological Change*, chs. 4 and 5.

18. Gunfounders: Barbiroli, *Repertorio storico*; Carpegna, *Brescian Firearms*; Morin and Held, *Beretta*. Local histories: Bolognini, 'La produzione e l'organizzazione del lavoro nelle fucine gardonesi'; Bossini and Galeri, *Valtrompia nella storia*; *Armi e cultura nel bresciano* (esp. the essay by Morin); *Antologia gardonese*; Odorici, *Cenni storici*; Cominazzi, *Cenni sulla fabbrica d'armi*.

19. Merlo, *Heavy Metal*; Mallett and Hale, *Military Organization of a Renaissance State*; Sherer, *Warriors for a Living*; Arfaioli, *Black Bands*; Mallett and Shaw, *Italian Wars*; C. Shaw, *Italy and the European Powers*; Pellegrini, *Le guerre d'Italia*; Bowd, *Renaissance Mass Murder*; Butters and Neher, *Warfare and Politics*.

20. Mocarelli and Ongaro, 'Weapons Production'; Ansani, 'Life of a Renaissance Gunmaker'; Panciera, 'Venetian Gunpowder'; Parrott, *Business of War*.

21. For an overview of the field, see C. Richardson, Hamling and Gaimster, *Routledge Handbook of Material Culture in Early Modern Europe.*

22. For example, Auslander and Zahra, *Objects of War.*

23. Bartels, 'Arms, Armour and Masculinity'; Neuschel, *Living by the Sword.*

24. Madden, *Civil Blood* makes the argument that increasing levels of vendetta violence in later sixteenth-century Italy likewise drove state formation; see, in particular, ch. 3.

25. Rogers, *Military Revolution Debate,* provides a compilation of key readings; see also Eltis, *Military Revolution.*

26. See, for example, chs. 3 (Rogers on the Hundred Years War) and 4 (Jeremy Black on the 1660–1792 period) in Rogers, *Military Revolution Debate,* and, on gunpowder weaponry, B. S. Hall, *Weapons and Warfare.*

27. Del Torre, *Venezia e la Terraferma,* 234.

28. Pezzolo, 'Republics and Principalities in Italy'; Pezzolo; 'La "rivoluzione militare"'. See also Capra, 'Italian States'; Caselli, 'Formation of Fiscal States in Italy'; Covini, 'Political and Military Bonds'; and Sherer, *Scramble for Italy,* 133.

29. Rogers, *Military Revolution Debate,* intro. and ch. 3.

30. Jacob and Visoni-Alonzo, *Military Revolution.*

31. P. H. Wilson and Klerk, 'Business of War Untangled'.

32. Fynn-Paul, *War, Entrepreneurs, and the State,* esp. intro.; Parrott, *Business of War.*

33. For discussions of costs in Spain and England respectively, see Thompson, '"Money, Money, and Yet More Money!"', 280; and G. Phillips, 'Longbow and Hackbutt', 589. More generally, see Eltis, *Military Revolution.* On the English gunpowder industry, see Cressy, *Saltpeter.*

34. Jacob and Visoni-Alonzo, *Military Revolution,* 4.

35. Kleinschmidt, 'Using the Gun'; McNeill, *Keeping Together in Time.*

36. Ruff, *Violence in Early Modern Europe*; Davies, *Aspects of Violence*; Blastenbrei, 'I romani tra violenza e giustizia'; Blastenbrei, 'Violence, Arms and Criminal Justice'; Dean and Lowe, *Murder in Renaissance Italy*; Rose, *Renaissance of Violence*; Cohn and Ricciardelli, *Culture of Violence.*

37. Davis, 'Up in Smoke'.

38. Important early works in this approach to the state were Elias, *Civilizing Process*; Oestreich, *Neostoicism and the Early Modern State*; and Foucault, *Discipline and Punish.* For subsequent application to sixteenth-century history, see Jütte, 'Poor Relief and Social Discipline'; Jütte, *Poverty and Deviance*; and Schilling, '"History of Crime" or "History of Sin"?'

39. Blockmans, Holenstein and Mathieu, *Empowering Interactions*; Karonen and Hakanen, *Personal Agency.*

40. Innes, 'Regulation of Charity', 20.

41. Barret, *Theorike and Practike,* 2.

42. LaTour, *Pandora's Hope,* 179.

43. Chase, *Firearms,* 199.

44. The difficulties are compounded by the fact that the Florentine military archive does not even have an index. Jacopo Pessina, *L'organizzazione militare.*

45. Archivio di Stato di Brescia (hereafter ASBs), Cancelleria prefettizia inferiore (hereafter CPI), Registri ducali 1 and 2; ASBo, Legato, Licenze di portare armi.

46. For example, Leonardo Mazzoldi's survey of the documentary sources for Gardone Val Trompia in *Antologia Gardonese*, 72–77, took a narrow view of what might be relevant and concluded there were minimal surviving sixteenth-century documents relating to the town.

47. De' Fedrici, in Ramusio, *Navigazioni*, 6:1062–63; De' Fedrici was there in 1569 although the account was not published until 1587 (pp. 1015, 1070).

Chapter 1

1. Sherer, *Warriors for a Living*, 220; Buchanan, '"Battle of Pavia"'; Le Gall, *L'honneur perdu*, 118–27.

2. Monluc, *Commentaires*, 1:52.

3. Nationalmuseum (Stockholm), inv. no. NM 272.

4. The relief showing the Battle of Ceresole victory can be viewed on Wikipedia Commons; see Racinaire, 'File:Scène de bataille'. The tomb was designed by Philibert de l'Orme and sculpted by François Camoy, François Marchand and Pierre Bontemps (esp. the last of these three). Hale, *Artists and Warfare*, 261.

5. Chase, *Firearms*, 32.

6. Chase, *Firearms*, 35; Andrade, *Gunpowder Age*, 112–13.

7. Ágoston, *Guns for the Sultan*, 15.

8. Chase, *Firearms*, 58–59, 78; Irwin, 'Gunpowder and Firearms', 119.

9. Chase, *Firearms*, 85; Irwin, 'Gunpowder and Firearms',121; Ágoston, *Guns for the Sultan*, 15–17.

10. Ágoston, *Guns for the Sultan*, 17.

11. Ágoston, *Guns for the Sultan*, 21.

12. Pepper and Adams, *Firearms and Fortifications*, 8, citing A. Hall, *Ballistics*, 9–14.

13. Pepper and Adams, *Firearms and Fortifications*, 8.

14. Ansani, 'Life of a Renaissance Gunmaker', 759.

15. Romanoni and Bargigia, 'La diffusione delle armi'.

16. Kluge and Dörschel, 'E fuero le prime'. The quotation is from the *Cronaca Senese*.

17. Tlusty, *Martial Ethic*, 19; Pius II, *Secret Memoirs*, 150.

18. For an overview of technological developments, see Morin, *Armi antiche*, 12–59.

19. Kunsthistorisches Museum (Vienna) (hereafter KHM), inv. no. A 65; see also A 66, A 70, and A 71 (larger arquebuses from a similar period).

20. Royal Armouries, inv. no. XII.5315; Mary Rose Museum, inv. no. 81A2679.

21. Dean, 'Eight Varieties', 86 and n. 23 (on 100).

22. On this and other *cassoni* of the period showing battles, see Nethersole, '*Armeggerie*'.

23. Mallett and Hale, *Military Organization*, 76, 141, n. 175.

24. Gelli, *Gli archibugiari*, 60, 62. For further background on the use of firearms in fifteenth-century Italy, see Mallett, *Mercenaries*, 156–59.

25. ASF, Otto di Pratica del Principato, 121.

26. Irwin, 'Gunpowder and Firearms', 124; the source is Spandugino, *Petit Traicté*, 108.

27. Ramusio, *Navigazioni*, 3:517.

28. La Rocca, 'Afonso "the African" and His Army', 38–40. I am grateful to Paulo M. Dias, Instituto de Estudos Medievais, Universidade NOVA de Lisboa, for drawing these tapestries to my attention.

29. Chase, *Firearms*, 78.

30. Interiano, 'Vita et Sito de Zychi', 169.

31. For further discussion, see Morin, 'Origins of the Wheellock', which takes issue with Blair, 'Further Notes on the Origin of the Wheellock'.

32. Mallett and Hale, *Military Organization*, 382.

33. Museu Militar de Lisboa, inv. no. MML00414; Real Armeria de Madrid, inv. no. K 53 (Morin, *Armi antiche*, cat. no. 35); Victoria and Albert Museum, inv. no. M 628-1927 (Morin, *Armi antiche*, cat. no. 50).

34. Morin, *Armi antiche*, cat. no. 35. For Peck's career and a catalogue of his known works, see Schalkhausser, 'Peter Peck'.

35. Greener, *Gun and Its Development*, 50.

36. Fitzwilliam Museum, inv. no. MAR.M 269-1912; Metropolitan Museum of Art, inv. no. L. 2018.47.

37. J. F. Hayward, *Art of the Gunmaker*, 1:58.

38. Krenn, Kalaus and Hall, 'Material Culture and Military History'.

39. Cellini, *Autobiography*, 87; *Vita*, 101.

40. Fourquevaux, *Instructions*, 98.

41. Cellini, *Autobiography*, 36–37; *Vita*, 47–48.

42. Arfaioli, *Black Bands*, 10–11, 19–20; Mallett and Shaw, *Italian Wars*, 143–44.

43. *Le armi degli Estensi*, 115.

44. Arfaioli, *Black Bands*; Mallett and Shaw, *Italian Wars*; C. Shaw, *Italy and the European Powers*; Pellegrini, *Guerre d'Italia*.

45. Pepper and Adams, *Firearms and Fortifications*, 11, citing Guicciardini, *History of Italy*, 1:148–49. For context, see Guilmartin, 'Military Revolution', 306–7.

46. Commynes, *Memoirs*, 2:199.

47. Irwin, 'Gunpowder and Firearms', 123.

48. Ágoston, 'War-winning Weapons?', 129–30, citing Pepper, 'Castles and Cannon'; see also Pepper, 'Face of the Siege'.

49. Chase, *Firearms*, 63–65; Lynn, '*Trace italienne*', 171–74; see also Pepper and Adams, *Firearms and Fortifications*.

50. Arnold, 'Fortifications', 222.

51. Leonardo, *Notebooks*, 276.

52. Veneranda Biblioteca e Pinacoteca Ambrosiana (Milan), Codex Atlanticus, fol. 33r.

53. Besides Figure 1.7, see also Codex Atlanticus, fols 113r and 977v.

54. Archivio di Stato di Venezia (hereafter ASVe), Capi del Consiglio dei Dieci (hereafter CCX), Dispacci dei Rettori (Brescia), 22, no. 74.

55. Museu Militar de Lisboa, inv. no. MML02473; Archivio Storico Capitolino (hereafter ASC), Fondo Orsini ser. II, 1205, under 'A' for 'Armeria'.

56. Leonardo, *Notebooks*, 237. For context, see Thommen, *Environmental History*, 122–23.

57. Leonardo, *Notebooks*, 236–37.

58. Leonardo, *Notebooks*, 237.

59. Leonardo, *Notebooks*, 237.

60. Erasmus, *Complaint of Peace*, 33.

61. Hale, *Artists and Warfare*, 131, fig. 179.

62. On this imagery, twelfth-century hostility to crossbows and bans on missile weapons more generally, see Strickland and Hardy, *Great Warbow*, 63, 116–19.

63. Gilino, *De Morbo*, 7. For context, see Arrizabalaga, Henderson and French, *Great Pox*, 50.

64. Strickland and Hardy, *Great Warbow*, 116–18.

65. Mallett and Hale, *Military Organization*, 79.

66. It may be possible to infer a link between the institution of the *camarada* in the Spanish army and its combat effectiveness; see Sherer, *Warriors*, 92–93, 241.

67. Mallett and Shaw, *Italian Wars*, 64–65.

68. 'Documenti per servire alla storia', 392, 429, 477; ASF, Nove conservatori di ordinanza e milizia, distribuzione di armi, 1.

69. Van Nimwegen, 'Transformation of Army Organisation', 162–67.

70. See also 'Armoured Soldiers Firing Match-Lock Arquebus'.

71. Bavarian State Library, Cod.icon. 222. For context, see Silver, *Marketing Maximilian*, esp. ch. 5, 147–68.

72. Archivio di Stato di Roma (hereafter ASR), Soldatesche e Galere (hereafter SG) 86 (Conti straordinari, 1478–1540), ins. 2.

73. ASVe, CCX, Dispacci dei Rettori (Brescia), 19, ins. 1507, no. 66.

74. Archivio di Stato di Modena (hereafter ASMo), Archivio Segreto Estense (hereafter ASE), Cancelleria ducale, Archivi militari 267, Libro di munizioni delle fortezze del duca di Ferrara, 1509.

75. ASR, SG 86 (Conti straordinari, 1478–1540), ins. 3, fols. 6r, 30v.

76. Strickland and Hardy, *Great Warbow*, 399, 405.

77. Royal Armouries, inv. no. XII.5315.

78. ASC, Fondo Orsini, ser. I, vol. 414/1, no. 85. For the conversions, see Cardarelli, *Encyclopaedia of Scientific Units*, 87.

79. Cesare d'Anselmi, 'Descrittione del sacco di Brescia', in *Sacco di Brescia*, 19–31 (here at 28). For a subsequent use of this text, see Luigi Da Porto, *Lettere storiche* in *Sacco di Brescia*, 177, and for the extract, 186.

80. Paredes, 'Confusion of the Battlefield'; Buchanan, '"Battle of Pavia"'.

81. Compare the sources collected in sections 4 (histories) and 5 (literary accounts) of *Sacco di Brescia* with those from local writers in sections 1 and 2 of the same volume.

82. Pandolfo Nassino, 'Registro di molte cose seguite', 139–57, in *Sacco di Brescia*, here at 139.

83. In *Sacco di Brescia*, 746.

84. Nassino, 'Registro di molte cose seguite', in *Sacco di Brescia*, 143, 146.

85. In *Sacco di Brescia*, 405–16. For background, see Piscini, 'Degli Agostini, Niccolò'.

86. *Sacco di Brescia*, editor's intro., 405.

87. In *Sacco di Brescia*, 406.

88. In *Sacco di Brescia*, 421.

89. Ascham, *English Works*, 49.

90. *Libro o vero Cronicha di tutte le guerre de Italia*, in *Sacco di Brescia*, 598.

91. Robert de Fleurange, 'Storia delle cose memorabili avvenute durante il regno di Luigi XII e di Francesco I 1525 circa', in *Sacco di Brescia*, 532; Jacques de Mailles, 'L'allegra storia del gentile signore di Bayard, 1527', in *Sacco di Brescia*, 540–52, here at 546.

92. ASVe, Patroni e Provveditori dell'Arsenal, 133, fol. 44v.

93. ASR, SG 86/4: Expenses of the 1517 Urbino campaign (Lorenzo de' Medici, duke of Urbino), pp. 7, 71–72. ASVe, CCX, Dispacci dei Rettori (Brescia), 20 (1538), no. 55.

94. ASMo, ASE, Cancelleria ducale, Archivi militari 269.

95. Francesco Guicciardini to Cardinal Giulio de' Medici, 31 August 1521, in Guicciardini, *Carteggi*, iv, 198–99; for context, see 186, 206. See also Mallett and Shaw, *Italian Wars*, 141.

96. Arfaioli, *Black Bands*, 10–11. Census: *Descriptio Urbis*, entry nos. 5525, 5528, 7020.

97. Arfaioli, *Black Bands*, 14.

98. Cellini, *Autobiography*, 91; *Vita*, 105.

99. Cellini, *Autobiography*, 60; *Vita* 72.

100. Cellini, *Autobiography*, 60; *Vita*, 72.

101. In *Sacco di Brescia*, 547.

102. Mexia, *Historia del emperador Carlos V*, 383, cited in Sherer, *Warriors*, 223.

103. Cited in Bowd, *Renaissance Mass Murder*, 80.

104. Guicciardini, *History of Italy*, 5:23.

105. Du Bellay, *Regrets*, sonnet 83. My translation.

106. Adapted from Cellini, *Autobiography* 66; *Vita*, 78. Bull's more elegant translation, 'My drawing, my wonderful studies, and my lovely music were all forgotten in the music of the guns', develops Cellini's metaphor beyond the original. I am grateful to Deanna Pellerano for drawing my attention to this point.

107. DeVries, 'Military Surgical Practice', 132–33.

108. Bowd, *Renaissance Mass Murder*, 125–26.

109. Nicolo Tartalea [Tartaglia], *Quesiti et inventioni diverse*; Tartaglia, *Three Bookes of Colloquies Concerning the Arte of Shooting*. On Tartaglia, see Pizzamiglio, *Atti della giornata di studio in memoria di Niccolò Tartaglia*.

110. In *Sacco di Brescia*, 165–68.

111. Brugh, *Gunpowder*, 52.

112. Brugh, *Gunpowder*, 88.

113. Schurink, 'War, What Is It Good For?', esp. 131–32 on Withorne's translation.

114. Machiavelli, *Art of War*, 96, 90.

115. Machiavelli, *Art of War*, 113.

116. ASC, Fondo Orsini, ser. I, vol. 412/1, ins. 21, fols. 1v and 2v.

117. *Nunziature*, 1:170–71.

118. ASF, Mediceo del Principato (hereafter MdP) 4849, fol. 303 (Medici Archive Project Document [hereafter MAP doc.] ID 22593). This is corroborated in Pessina, *L'organizzazione militare*, 129.

119. ASMo, ASE, Cancelleria ducale, Archivi militari 45. By comparison, a 1571 muster of Spanish troops in the Netherlands included about 30 per cent arquebusiers. Parker, '"Military Revolution"', 50, n. 11.

120. Archivio di Stato di Napoli (hereafter ASNa), Archivio Farnese 573 (Castro e Ronciglione), fol. 30v (here referring to stamped foliation).

121. Quarenghi, *Tecno-cronografia*, 1:239.

122. Mora, *Il Soldato*, 88, cited in Arfaioli, *Black Bands*, 20.

123. ASVe, CCX, Dispacci dei Rettori (Brescia), 21, no. 192.

124. Brugh, *Gunpowder*, 64.

125. Ramusio, *Navigazioni*, 3:695.

126. Garimberto, *Capitano Generale*, 494.

127. Guicciardini, *History of Italy*, 1:149.

Chapter 2

1. Mallett and Shaw, *Italian Wars*, 209.

2. A recent overview with bibliography is in Zan, *Venice Arsenal*, 11–21.

3. Sabatti, 'La Valtrompia nel '500', 165; and see above, note 1 to chapter 1.

4. B. Richardson, *Printing, Writers and Readers*, 25–35.

5. ASBs, Archivio Notarile, b. 1965.

6. The collections of Brescia Museums include various fragments dating from the seventh to the third century BCE, most now on display at Museo ORMA in Pezzaze; see, for example, inv. nos. ST 66899–900, 66901A–B, 66902A–F, 66904–6, 66907A–B.

7. Marchesi, 'Le miniere', 29.

8. Sanudo, *Itinerario*, 288–89; his reference to the valleys is at 280–81.

9. Leonardo da Vinci, 'Three Sketches of the Course of the Rivers Brembana, Trompia and Sabbia', Royal Collection, inv. no. RCIN 912673.

10. Morin and Held, *Beretta*, 24. For further background on early production in Gardone, see Riccadonna, 'Dal minerale di ferro all'arcobugio'; and Belfanti, 'Chain of Skills'.

11. Bolognini, 'La produzione', 16–19; Morin, 'La produzione', 70–71.

12. Sabatti, 'Valtrompia', 165, citing the 1527 relation to the Senate of Antonio Tiepolo, podestà of Brescia, in Tagliaferri, *Relazioni*, 16–17.

13. Ferraro, *Family and Public Life*, 155–56; Del Torre, *Venezia e la Terraferma*, ch. 8.

14. Tagliaferri, *Relazioni*, 79; Sabatti, 'Valtrompia', 179. A report of 1520 estimated the population to be fifty thousand (Sabatti, 'Valtrompia', 163, citing Tagliaferri, *Relazioni*, 3): this may have reflected numbers who had fled to the valleys during the French occupation.

15. For some examples see ASVe, CCX, Dispacci dei Rettori (Brescia), 21, nos. 171–82.

16. ASVe, CCX, Dispacci Rettori (Brescia), 19, no. 50; discussed in Morin and Held, *Beretta*, 24. The letter does not give detail of the questioning.

17. McCray, *Glassmaking*, 130–36.

18. ASVe, CCX, Dispacci dei Rettori (Brescia), 19, no. 59. Occasional further references to such purchases continue in the documents of the subsequent decade. See, for example, ASVe, CCX, Dispacci dei Rettori (Brescia), 19, ins. 1516, no. 73, 29 September, a letter from Andrea Trevisan with reference to ordering *ballotte* from Valtrompia.

19. *Antologia gardonese*, 172; Sabatti, 'Valtrompia,' 160, n. 26, notes there is no source for this claim, but it is repeated in Museo delle Armi e della Tradizione Armeria di Gardone Valtrompia, 'Lavorazione Artigianale'.

20. Sabatti, 'Valtrompia', 159.

21. Sabatti, 'Valtrompia', 163 (population) and 165 (food supplies), citing Tagliaferri, *Relazioni*, 3 and 16–17.

22. A report of 1801 gives the date 1619. Belfanti, 'Chain of Skills', 277.

23. Biringuccio, *Pirotechnia*, 19r.

24. On the painting and its sources, see the catalogue entry by Luc Serck in *Autour de Henri Bles*, 242–43.

25. Pessina, *L'organizzazione militare*, 131–35.

26. Sabatti, 'Valtrompia', 160; Pepper, 'Patriots', 96–97; Francesco Gambara, 'Ragionamenti di cose patrie ad uso della gioventù', in *Sacco di Brescia*, 53–54.

27. Gian Giacomo Martinengo di Erbusco, 'Della congiura de' Bresciani per sottrarre la patria alla francese dominazione', in *Sacco di Brescia*, 61–118, here at 70.

28. *Annali di Bovegno*, 131.

29. *Annali di Bovegno*, 131.

30. For a brief narrative of events, see Bowd, *Venice's Most Loyal City*, 214–16.

31. See, for example, the letter of 5 June 1528: ASVe, CCX, Dispacci dei Rettori (Brescia), 19, no. 218.

32. For the Chino/Chini family's presence in Gardone in the first half of the sixteenth century, see also the 1509 contract signed in the house of Giovanni de' Chini: ASBs, Archivio Notarile, b. 1965, unnumbered contract.

33. ASVe, CCX, Dispacci dei Rettori (Brescia), 19, no. 298. Tagliaferri, *Relazioni*, liii, standardizes the captain's name as Giacomo Correr.

34. On contracting practices in early modern Venice, see J. E. Shaw, 'Informal Economy'.

35. ASVe, CCX, Dispacci dei Rettori (Brescia), 20, nos. 23, 27, 30.

36. ASVe, CCX, Dispacci dei Rettori (Brescia), 20, no. 50.

37. Tagliaferri, *Relazioni*, 79–80.

38. Tagliaferri, *Relazioni*, 48.

39. Tagliaferri, *Relazioni*, 117.

40. 'Relazione di Domenico Priuli', in Tagliaferri, *Relazioni*, 115–50 (quotation at 117).

41. For the continuation of this problem into the seventeenth century, see Mocarelli and Ongaro, 'Weapons Production'.

42. ASBs, CPI, Registri ducali 2, fol. 44v.

43. For examples of the privileges, see ASBs, CPI, Registri ducali 1, fols. 91v–92r, 103r, 146v, 202r–v, 205r–v.

44. ASBs, CPI, Registri ducali 1, fol. 91–92r.

45. ASBs, CPI, Registri ducali 1, fol. 205r–v.

46. ASBs, CPI, Registri ducali 2, fol. 95r.

47. Sabatti, 'Valtrompia', 160; Pepper, 'Patriots and Partisans', 79–103.

48. Ferraro, *Family and Public Life*, 13; Montanari, *Quelle terre*, esp. pt. 2, ch. 2, 'Valcamonica. Una valle alle frontiere di San Marco', 161–83.

49. Sanudo, *Itinerario*, 278, and commentary, 279; Montanari, *Quelle terre*, 187–202, includes transcripts of four privileges spanning 1440–48. On Brescia, its patriciate and relations with Venice, see Bowd, *Venice's Most Loyal City*, esp. chs 3 and 4; Ferraro, 'Feudal-Patrician Investments', 32; Montanari, *Quelle terre*; Valseriati, *Tra Venezia e l'Impero*; and Del Torre, *Venezia e la Terraferma*.

50. I discuss these privileges in greater detail in Fletcher, 'Venice, Brescia and Gardone Val Trompia'.

51. Tagliaferri, *Relazioni*, 49, 34.

52. ASBs, CPI, Registri ducali 1, fol. 146v. A letter on this subject from the rectors from May 1531 is in ASVe, CCX, Dispacci dei Rettori (Brescia) 19, no. 257.

53. ASBs, CPI, Registri ducali 1, fol. 202r–v.

54. ASVe, CCX, Dispacci dei Rettori (Brescia) 19, no. 311.

55. Further 1533 correspondence between the rectors and Venice regarding valley matters is in ASVe, CCX, Dispacci dei Rettori (Brescia) 19, nos. 283, 303. For 1548 and 1549, see ASBs, CPI, Registri ducali 2, fols. 79v, 96r–v.

56. Pessina, *L'organizzazione militare*, 195, citing a report of the captain Giovan Francesco Maraffi.

57. *Documenti inediti*, 579–80.

58. ASBs, CPI, Registri ducali 2, fol. 77v.

59. ASBs, CPI, Registri ducali 3, fol. 126r.

60. ASVe, CCX, Dispacci dei Rettori (Brescia), 21, no. 157; ASBs, CPI, Registri ducali 3, fol. 140r.

61. Targioni-Tozzetti, *Notizie*, 233; ASF, MdP 638, fol. 353r (MAP doc. ID 15371).

62. *Documenti inediti*, 587–88.

63. Targioni-Tozzetti, *Notizie*, 234.

64. ASF, MdP, 221, fol. 53r (MAP doc. no. 4856).

65. De' Lancellotti, *Cronaca*, 33–34. On Alfonso, see below, chapter 8.

66. Bertolotti, 'Le arti minori', 541.

67. ASC, Fondo Orsini, ser. I, vol. 414/1, ins. 119, fol. 130v.

68. Zanelli, 'Un contratto'. Spanish-style guns were in the Orsini armoury in 1568: ASC, Fondo Orsini, ser. I, vol. 412/1, ins. 4, fol. 1665r.

69. Royal Armouries, inv. no. XII.5315; Mary Rose Museum, inv. no. 81A2679.

70. ASR, SG 89/unnumbered: Conti di Niccolò Ultramonte da Bellano proveditor della Fonderia de N. S. dal 1562 al 1565, fol. 11v.

71. The 1612 edition of the *Vocabolario della Crusca* says that strictly speaking a *soma* is the load that may be borne by a beast of burden. Under the metric system, *soma* was specified as a hectolitre (100 litres).

72. ASBs, CPI, Registri ducali 1, fol. 17r

73. ASBs, CPI, Registri ducali 1, fol. 33v.

74. ASVe, CCX, Dispacci dei Rettori (Brescia) 22, no. 126.

75. Mallett and Hale, *Military Organization*, 353. On the earlier use of firearms by more irregular Venetian forces, see Pepper, 'Patriots'.

76. ASBs, CPI, Registri ducali 1, fol. 43r.

77. ASBs, CPI, Registri ducali 1, fol. 29r.

78. ASBs, CPI, Registri ducali 1, fol. 33v.

79. ASBs, CPI, Registri ducali 1, fol. 47v.

80. ASBs, CPI, Registri ducali 1, fol. 184r–v. There may be a seventeenth consignment listed on a damaged part of the MS.

81. ASBs, CPI, Registri ducali 1, fol. 195v.

82. ASMo, ASE, Camera, Amministrazione della casa, Armeria, b. 1, carteggio diverso, unnumbered letter.

83. ASVe, CCX, Dispacci dei Rettori (Brescia) 21, no. 186.

84. ASMo, ASE, Camera, Amministrazione della casa, Armeria, b. 1, carteggio diverso, unnumbered letter.

85. *Nunziature*, 2:284.

86. *Nunziature*, 2:292, 298.

87. Morin and Held, *Beretta*, 44, citing ASVe, Collegio, Relazioni, b. 37, fol. 4r–v, published in Tagliaferri, *Relazioni*, 40–41.

88. *Nunziature*, 1:141–42.

89. *Nunziature*, 1:77.

90. ASVe, CCX, Dispacci dei Rettori (Brescia) 20, no. 53.

91. *Nunziature*, 2:280–81.

92. *Nunziature*, 5:109.

93. *Nunziature*, 5:122 n. For further background, see Guerrini, *Congregazione dei padri della pace*, 89–92.

94. *Nunziature*, 5:111, 116.

95. *Nunziature*, 5:126.

96. *Nunziature*, 5:198.

97. For the issue arising in September 1551, see *Nunziature*, 5:280–81.

98. *Nunziature*, 5:318. For an example of the tensions, see ASVe, CCX, Dispacci dei Rettori (Brescia), 22, nos. 153–54 and 170–74.

99. *Nunziature*, 8:241–42.

100. *Nunziature*, 8:242, 249.

101. Klötzer, 'Melchiorites and Münster', 236, 241; Goertz, *Anabaptists*, 104.

102. ASVe, CCX, Dispacci dei Rettori (Brescia), 22, nos. 107, 116.

103. ASNa, Archivio Farnese, Parma, 398/2, fol. 9r. On the wider politics, see Brigden, 'Crusade'.

104. Bowd, *Reform before the Reformation*, 215.

Chapter Three

1 . ASBs, Archivo Martinengo dalle Palle, b. 76, 'Marcantonio Porcellaga, 1539-1559' (hereafter Porcellaga). The book was maintained from 1539 to 1559 by more than one member of the family. Although the binding is labelled 'Marcantonio Porcellaga', the contents are written in more than one hand, and the text repeatedly documents transactions made by Giovanni Battista Porcellaga, as well as referring on multiple occasions to a son named Scipione; Marcantonio, another of Giovanni Battista's sons, probably took over after the death of his father. On the family, see Pasero, *La partecipazione bresciana*, 8, 9, 16, 28, 32, 71 n. 80. The Porcellaga account books are in the archive of another noble family, the Martinengo. However, the absence of internal reference to the Martinengo in the books, along with references by Porcellaga to conducting transactions in his own house, suggest a later confluence of archives rather than a direct role for Porcellaga in Martinengo service.

2. Fynn-Paul, *War, Entrepreneurs, and the State*, esp. the introduction by Fynn-Paul, Marjolein 't Hart and Griet Vermeesch, 1–12; Parrott, *Business of War*; P. H. Wilson and Klerk, 'Business of War Untangled'.

3. For a recent exception concerning eighteenth-century Italy, see Martoccio, 'Place for Such Business'.

4. For more on the figure of the arms dealer in historiography, see Fletcher, 'Worst Business'.

5. On mercenaries and the contracting system, see Parrott, *Business of War*; and Mallett, *Mercenaries*.

6. His date of birth is given in the genealogical tables of the Società Storica Lombarda without a source, but 1492 seems plausible given that his son Scipione was of age to take on a substantial military contract in 1549. See Società Storica Lombarda, 'Giovanni Battista Porcellaga + Cecilia Gavardo'.

7. ASBs, CPI, Registri ducali 3, fol. 14v. On the Brescian elite, see Ferraro, *Family and Public Life*, ch. 2, 51–71. For Porcellaga's role in Asola, see the correspondence in ASBs, Archivio Martinengo dalle Palle, b. 91.

8. Valseriati, *Tra Venezia e l'Impero*, 64. For further accusations of involvement in violence, see ASBs, Archivio Martinengo dalle Palle, b. 91.

9. For a list of *condotte* held by the Orsini between 1469 and 1533, see C. Shaw, *Political Role of the Orsini*, app.

10. Brunelli, 'Camillo Orsini', gives a biographical overview.

11. See Mallett, *Mercenaries*, 121, 237, 252–54; and Damiani, 'Sforza Pallavicini'.

12 Porcellaga, fol. 42v.

13. Fantoni, *La Corte del Granduca*, esp. sec. III.a: 'Il dono: liberalità e potere', 97–137.

14. Porcellaga, fols. 77r, 111v.

15. Porcellaga, fol. 112v.

16. Fletcher, *Diplomacy*, ch. 7.

17. Porcellaga, fol. 120r, 123v.

18. Porcellaga, fols. 128v, 129r.

19. Porcellaga, fol. 169r.

20. Porcellaga, fol. 41v.

21. Porcellaga, fols. 42v, 114v, 142v.

22. Porcellaga, fol. 52r.

23. ASBs, Archivio Martinengo dalle Palle, b. 91, fol. 61r: 'subbito gionto messer Gio: Finardo da Bergomo lo haggio agrezato quanto ho possuto cosi i patroni delle arme à mandarli subbito, et venir lui dreto con esse'.

24. ASBs, Archivio Martinengo dalle Palle, b. 91, fol. 5v, letter of 4 November 1551.

25. Porcellaga, fols. 79r, 92r.

26. ASVe, CCX, Dispacci dei Rettori (Brescia), 21, nos. 42–43.

27. Archivio Storico Civico di Brescia (ASCBs), Archivio Gambara 609, Registro delle spese di casa di Lucrezio e Nicolò Gambara, fols. 2v, 8r, 17r, 19v, 32v, 35r, 39r. The Venetian transaction, for Zaccaria Barbaro, is in ASCBs, AG 608, unpaginated insert.

28. ASCBs, Archivio Gambara 609, Registro delle spese di casa di Lucrezio e Nicolò Gambara, fol. 18v.

29. On the diplomatic use of hospitality, see Fletcher, '"Furnished with Gentlemen"'.

30. ASBs, CPI, Registri ducali 2, fol. 44v, Pontificio Stato, export licence, including list of subsequent separate consignments, 28 February 1547; fol. 46v, Pontificio Stato, export licence, and consignment details (though not for arquebuses) 9 March 1548.

31. Porcellaga, fol. 113v.

32. Porcellaga, fol. 125v.

33. Ferraro, *Family and Public Life*, 70.

34. ASVe, CCX, Dispacci dei Rettori (Brescia), 19, no. 49, letter of 1 April 1505; fol. 59r, letter of 11 October 1506. The Negroboni relationship with the Brescian rectors and Venice continued over decades; see ASVe, CCX, Dispacci dei Rettori (Brescia), 21, nos. 18, 163, 164; and *Sacco di Brescia*, 499.

35. ASF, Guardaroba Medicea (hereafter GM) 31, fol. 69: 'per campione di piu somma che si ha a fare'.

36. ASVe, CCX, Dispacci dei Rettori (Brescia) 19, no. 311; 20, no. 8; Pessina, *L'organizzazione militare*, 135–36; ASVe, CCX, Dispacci dei Rettori (Brescia), 22, no. 3.

37. Early modern Italian measurements often varied regionally, and it is hard to provide precise equivalents. A *braccio* (plural *braccia*) was an arm's length, ranging between 58.3 cm in Florence and 68.3 cm in Milan. A *lira* or *libbra* was a measure of weight, around 300 g or a little more. *Onze* were used to measure both length and

weight, representing one twelfth of a *braccio* or *libbra*. Cardarelli, *Encyclopaedia*, 87–88.

38. Zanelli 'Un contratto'. Quarenghi, *Tecno-cronografia*, 1:184, cites Cominazzi, *Cenni*, but does not provide an archive source for the additional contract text presented.

39. Porcellaga, 150v.

40. Porcellaga, 144r.

41. ASBs, CPI, Registri ducali 2, fol. 44v, Pontificio Stato, export licence, including list of subsequent separate consignments, 28 February 1547; fol. 46v, Pontificio Stato, export licence, and consignment details (though not for arquebuses), 9 March 1548.

42. ASR, Soldatesche e Galere (hereafter SG) 88, volume entitled 'Conto delli denari che si spenderanno nelle cose della militia per presidio et securezza dell'Alma Città di Roma . . .', unpaginated entry in *Uscita* for 15 December 1552. Further discussion of these accounts is in Fletcher, 'Agents'.

43. See Ferraro, *Family and Public Life*, 70, for the Porcellaga as old nobility, and 76 for their participation in the council between 1588 and 1650.

44. Ferraro, *Family and Public Life*, 61.

45. Porcellaga, fol. 127v.

46. Pasero, *Partecipazione bresciana*, 32.

47. Pasero, *La partecipazione bresciana*, 71, n. 80. Pasero does not give a date for the grant of this title.

48. Zanelli, 'Un contratto'.

49. Bertolotti, 'Le arti minori', 584; ASF, MdP 1850, fol. 612r (MAP Doc. ID 20156).

50. ASBs, CPI, Registri ducali 1, fols. 150v–151r.

51. ASR, SG 646 (Miscellanea di carte sciolte), 'Libro di spese fatte per M. Orlando Riccio', 1541–42, fol. 9v.

52. ASR, SG 88, ins. 3, 'Armi distribuite e vendute alle comunità Stato Ecclesiastico nel 1550'. For the use of mules in transporting munitions, see also ASC, Orsini, ser. I, 294 no. 127.

53. ASR, SG 90, unnumbered insert.

54. ASR, SG 4, ins. 3, 'Conto d'arme ricevute e distribuite alle battaglie di Nostro Signore de la Reverenda Camera 1575'.

55. See, for example, the 1544 inventories of fortresses in ASMo, ASE, Cancelleria ducale, Archivi militari 271.

56. ASR, SG 88, ins. 3.

57. ASF, Otto di Pratica del Principato, 121, 'Distributione fatta per li Spettabili Nove' (1530); ASMo, ASE, Cancelleria ducale, Camera, Amministrazione della casa, Armeria, b. 1, reg. 5: 'Nota delle persone che dicono sono provviste delle armi a loro tassate' (1552).

58. ASBs, CPI, Registri ducali 1, fol. 161v.

59. ASR, SG 4, ins. 2, inventory of the fortress at Perugia, unpaginated.

60. See for example ASC, Orsini, ser. II, 1806, fol. 38r.

61. Panciera, 'Venetian Gunpowder', 106; ASVe, CCX, Dispacci dei Rettori (Brescia), 20, no. 112.

62. On the technology of gunpowder production, and the circulation of Italian knowledge, see Cressy, *Saltpeter*, 11–25.

63. ASVe, CCX, Notatorio, Registri 7, fol. 98r.

64. ASC, Fondo Orsini, ser. I, vol. 414/1, no. 118, fols. 121r and 122r.

65. ASC, Fondo Orsini, ser. II, 1806, fol. 103v.

66. ASR, SG 646 (Miscellanea di carte sciolte), unpaginated accounts for 1575.

67. ASR, SG 646, 'Libro di spese fatte per M. Orlando Riccio', fol. 9v.

68. ASR, SG 87, 'Conto di riscossioni e pagamenti per l'armata di S. S., 1538–39', fols. 50r, 54r.

69. ASR, SG 88 (Conti Straordinari, 1541–52), ins. 7 (Guerra di Paliano), pp. xl and 41; ASR, SG 646 'Libro di spese fatte per M. Orlando Riccio', fol. 18r.

70. ASC, Fondo Orsini, ser. I, vol. 414/1, no. 113: inventory of 22 January 1522 of Rocca di Palo.

71. ASC, Fondo Orsini, ser. I, vol. 414/1, no. 118, fol. 122r.

72. ASMo, ASE, Cancelleria, Archivi Militari, 139, unnumbered 'Relacione delle municioni da Guerra'.

73. ASF, Nove conservatori di ordinanza e milizia, Entrata e uscita / Debitori e creditori, 19, p. 44.

74. ASNa, Archivio Farnese, 573, fol. 8r.

75. Roberts, 'Military Revolution', 15.

76. Morin, 'Origins', 88, citing ASVe, CCX, Parti Comuni, reg. 18, fol. 170r; ASBo, Legato, Bandi 1, fols. 340r–341v. R. A. Kea observes that blacksmiths on the West African coast acquired the skills to repair and improve firearms: 'Firearms and Warfare', 205.

77. ASCBs, Archivio Gambara, 610, unpaginated notebook of 'Dinari recevuti per me Cesare dal Signor Conte'.

78. ASBo, Tribunale del Torrone, Registri (1560–1602), b. 1134, fol. 79v.

79. ASC, Orsini, ser. I, vol. 414/1, no. 81.

80. In *Sacco di Brescia*, 166.

81. ASR, SG 88, ins. 3, 'Armi distribuite', fol. 12v. 'Burnishing' was also a role undertaken in the workshop of Benedetto Spadaro: see Porcellaga, fol. 94r.

82. Belfanti, 'Chain of Skills', 268, citing ASBs, Curia prefettizia superiore, b. 27, fol. 19.

83. ASBo, Legato, Bandi 1, fols. 242v–243r, *bando* (proclamation) of 16 October 1555: 'alcuni hanno trovato modo d'usare gli Archibugietti piccioli, senza Ruota, portandoli coperti in diversi modi'.

84. ASBs, Archivio Martinengo dalle Palle, b. 76, 'Spese di Gio. [Battista] Porcellaga dal 1498 fino al 1539', fol. 76v.

Chapter Four

1. Bertolotti, 'Le arti minori', 583.

2. Hale, 'Gunpowder'; and see above, chapter 1.

3. Cellini, *Autobiography*, 141; *Vita*, 154.

4. Cellini, *Autobiography*, vii (intro. by George Bull).

5. ASF, GM 30, fol. cv (this volume uses Roman numerals for recto folios and Arabic for verso).

6. Alessandro: Fletcher, *Black Prince*, 33; Paolo Giordano: ASC, Fondo Orsini, ser. I, vol. 414/1, no. 79 (1 October 1553).

7. Castiglione, *Courtier*, 57.

8. The classic work is Elias, *Civilizing Process*; on its limits, see Rose, *Renaissance of Violence*, 3.

9. Fantoni, 'Un rinascimento a metà'.

10. Mann, 'Lost Armoury I', 256.

11. *Le armi degli Estensi*, 18.

12. This is particularly the case for smaller weapons: the larger and heavier guns known as *archibusoni*, which had to be fired from a rest, were generally recorded alongside artillery.

13. Hale, 'Gunpowder', 132.

14. Hetoum, *Lytell cronycle*, 8, cited in Chase, *Firearms*, 59.

15. Brugh, *Gunpowder*, 7.

16. Brugh, *Gunpowder*, 55.

17. Brugh, *Gunpowder*, 75.

18. Fronsperger, *Kriegsbuch*, 1:lxxxviiia, cited in Brugh, *Gunpowder*, 81; for context, see also 79–80.

19. Ariosto, *Orlando Furioso*, canto 11, verse 26, cited in Hale, 'Gunpowder', 121.

20. Fronsperger, *Kriegsbuch*, 1:clxxiib, cited in Brugh, *Gunpowder*, 76.

21. Fronsperger, *Kriegsbuch*, 1:clxxiia, cited in Brugh, *Gunpowder*, 77.

22. Montaigne, *Essays* (1885), 144, cited in Chilton, 'Humanism and War', 134.

23. Chilton, 'Humanism and War', 137.

24. 'Hollowed bronze' appears in an anonymous poem; see *Battle of Lepanto*, 102–3.

25. In *Sacco di Brescia*, 442.

26. Gambara, *Complete Poems*, 133, cited in Bowd, *Renaissance Mass Murder*, 213.

27. Cellini, *Autobiography*, 41; *Vita*, 53.

28. Cellini, *Autobiography*, 42; *Vita*, 54.

29. Stapleford, *Lorenzo de' Medici at Home*, 38, 192. See also Scalini, 'Weapons of Lorenzo de' Medici'.

30. ASF, GM 7 (Inventario della Guardaroba del Duca Cosimo alla consegna di Giovanni Ricci da Prato), fol. 47r.

31. ASMo, ASE, Cancelleria ducale, Camera, Amministrazione della casa, Armeria, b. 1, reg. 3, fols. 7v, 21v, 36v.

32. ASMo, ASE, Cancelleria ducale, Camera, Amministrazione della casa, Armeria, b. 1, regg. 1 and 2.

33. ASMo, ASE, Cancelleria ducale, Camera, Amministrazione della casa, Armeria, b. 1, reg. 4.

34. Archivio di Stato di Parma (hereafter ASP), Casa e corte farnesiana, b. 52 (Inventari), fasc. 1; 1540 inventory of Mastro di camera.

35. ASP, Casa e corte farnesiana, b. 52 (Inventari), fasc. 1; 1540 inventory of Mastro di camera.

36. ASP, Casa e corte farnesiana, b. 52 (Inventari), fasc. 1; 1545 inventory of Ottavio Farnese.

37. 'Archibusi vechij corti deti pistoni no. 7.' ASP, Casa e corte farnesiana, b. 52 (Inventari), fasc. 1; Fortezza di Felino.

38. Mann, 'Lost Armoury I', 298–99, 312–13, 296–97, 308–9 (inv. nos. 187, 348, 165 and 287).

39. Mann, 'Lost Armoury II', 116–17 (inv. no. 560).

40. Foley et al., 'Wheellock', 415.

41. Veneranda Biblioteca e Pinacoteca Ambrosiana (Milan), Codex Atlanticus, fol. 987r.

42. Cardinal Sforza: Gaibi, *Armi da fuoco*, 21, citing a poem by Cardinal Castellesi to this effect. Cardinal d'Este: Morin, 'Origins of the Wheellock', 84–85; Franci, *Wheellock Firearms*, 6; Brooker, 'What Can Be Learned', 9.

43. Royal Armouries, inv. no. 40782.

44. Brooker, 'What Can Be Learned'.

45. ASF, GM 7 (Inventario della Guardaroba del Duca Cosimo alla consegna di Giovanni Ricci da Prato), fol. 47r.

46. ASBs, CPI, Registri ducali 2, fol. 189r. A similar privilege was extended to cavalry.

47. *Documenti inediti*, 324.

48. Morin, 'Origins', 88, citing ASVe, CCX, Parti Comuni, reg. 23, fol. 155r–v.

49. Schwoerer, *Gun Culture*, 126.

50. In Bayerisches Nationalmuseum (Munich), inv. no. W 1498; Morin, *Armi antiche*, cat. no. 26; and Real Armeria de Madrid, inv. nos. K32 and K33. Franci, *Wheellock Firearms*, 6.

51. Bercusson, 'Gift-giving, Consumption and the Female Court'. On Johanna's involvement in arms procurement, see Arrivo, 'Una dinastia', 54.

52. Bertolotti, 'Le arti minori', 584.

53. ASMo, ASE, Camera, Amministrazione della casa, Armeria, b. 1, reg. 4; ASP, Casa e corte farnesiana, b. 52 (inventories), fasc. 1; 1540 inventory of Mastro di camera.

54. ASF, GM 43, p. 23.

55. Metropolitan Museum of Art, inv. no. 14.25.1433a, b, 1523.

56. Mann, 'Lost Armoury I', 312–13 (no. 348 of 1542 inventory); 'Lost Armoury II', 58–59 (nos. 92 and 93 of Libro di Aquila).

57. Mann, 'Lost Armoury II', 118–19 (no. 574 of Libro di Aquila).

58. Mann, 'Lost Armoury II', 112–13 (no. 542 of Libro di Aquila); see also nos. 555 and 563. For the crossbows, see Mann, 'Lost Armoury II', 72–73 (nos. 209, 212) and discussion at 18 and 25–26.

59. ASF, MdP 2, fol. 125r (MAP doc. ID 19990).

60. ASF, MdP 1852, fol. 574r (MAP doc. ID 21465).

61. ASF, GM 30, fol. xcviij. For Centurione, see Nuti, 'Adamo Centurione'.

62. ASF, GM 30, fol. c. For Balena's role, see fol. cv. On the cardinal's hunting, see *State Papers*, 7:150 (*Letters and Papers*, 4: no. 5213).

63. ASC, Orsini, ser. II, 1205, unnumbered inventory of Paolo Giordano Orsini. The archive dates this to 1578–81.

64. Metropolitan Museum of Art, inv. no. L.2017.9.2a-k (on loan); Ágoston, *Guns for the Sultan*, 23–24; *Süleymanname*, 118–23.

65. On the Renaissance home and its decorative style, see P. Thornton, *Italian Renaissance Interior*; Ajmar-Wollheim and Dennis, *At Home*.

66. Franci, *Wheellock Firearms*, 13.

67. Springer, *Armour and Masculinity*, 6.

68. Metropolitan Museum of Art, inv. no. 14.25.1493.

69. Metropolitan Museum of Art, inv. no. 29.151.1.

70. An image of Lucretia and nude allegorical images of justice and love feature on a German wheellock rifle of c. 1580–90; see Wallace Collection, inv. no. A1083. For the Trojan War see Art Institute of Chicago, inv. no. 1982.2266.

71. Lucretia and Judith: Wallace Collection, inv. no. A1075; Venus: KHM, Hofjagd- und Rüstkammer, inv. no. A 2305.

72. Carpegna, *Antiche armi*, 76: cat. no. 466: 'Pistola a ruota, Germania 1548, Inv. N. 1518'. J. F. Hayward, *Art of the Gunmaker*, 1:280, pl. 6 b; Morin, *Armi Antiche*, cat. no. 41.

73. ASF, GM 43, p. 23.

74. *Le armi degli Estensi*, 115, Konopiště Castle, inv. no. 66.

75. ASF, GM 43, p. 23.

76. Morin, *Armi antiche*, cat. no. 30 (Metropolitan Museum of Art, inv. no. 14.25 1425). See also Schalkhausser, 'Peter Peck', 191.

77. Statens Historiska Museer, Livsrustkammer (Stockholm), inv. no. 16318 LRK.

78. ASC, Orsini, ser. II, 1205.

79. Museo Correr (Venice), inv. no. Cl. XIV n. 0356.

80. Morin, *Armi antiche*, cat. no. 34. For a further example see the 1550 double-barrelled wheellock in the KHM, Hofjagd- und Rüstkammer, inv. no. A 518. Flasks: Metropolitan Museum of Art, inv. no. 14.25.1491 / 2014.191.

81. ASF, GM 30, fol. ciij.

82. ASF, GM 30, fol. cv.

83. P. Thornton, *Italian Renaissance Interior*, 92–94; Ajmar-Wollheim and Dennis, *At Home*, 120–21 ('Chests' by Claudio Paolini), and 224–25 ('Tables and Chairs' by Fausto Calderai and Simone Chiarugi).

84. P. Thornton, *Italian Renaissance Interior*, 100.

85. P. Thornton, *Italian Renaissance Interior*, 85–86.

86. Sicca, 'Consumption and Trade of Art', 169.

87. KHM, Hofjagd- und Rüstkammer, inv. no. A 2305.

88. Art Institute of Chicago, inv. no. 1982.2266.

89. See the examples indexed in Ajmar and Wollheim, *At Home*.

90. ASF, GM 30, fol. c.

91. ASF, GM 45 (1560), fol. 53r.

92. ASF, GM 44 (Inventario a capi), fols. 53r, 54r.

93. M. Hayward, *Dress at the Court of Henry VIII*, 20.

94. Knecht, *Renaissance Warrior*, 131.

95. Steadman, *Renaissance Fun*, 165–66.

96. Ajmar-Wollheim and Dennis, *At Home*, 214 (essay by Ajmar-Wollheim, 'Sociability', 206–21).

97. *Armi degli Estensi*, 114 (Konopiště Castle inv. no. 65); for a gun with similar mixed mechanisms, see inv. 34.60 at the Metropolitan Museum of Art.

98. J. F. Hayward, *Art of the Gunmaker*, 1:59 (Danner), 69 (Peck). For Peck, see also Schalkhausser, 'Peter Peck'; Morin, *Armi antiche*, includes multiple examples of guns by Peck; for Danner, see also Morin, *Armi antiche*, cat. no. 44 (Statens Historiska Museer, Livsrustkammer (Stockholm), inv. no. 16318 LRK).

99. Morin, *Armi antiche*, cat. nos. 9 (Le Sale d'armi del Consiglio dei Dieci nel Palazzo Ducale [Venice], inv. 2031/30), 10 (Museo Civico Luigi Marzoli [Brescia], inv. no. 800) and 48 (Armeria Reale [Turin], inv. no. 49).

100. ASF, GM 30, fol. cv (105 bis).

101. ASF, GM 45 (1560), fol. 53r.

102. Royal Armouries, inv. no. XIV.5.

103. For full details of Henry's gun ownership, see *Inventory of King Henry VIII*.

104. Morin, *Armi antiche*, cat. no. 27 (Le Sale d'armi del Consiglio dei Dieci nel Palazzo Ducale [Venice], inv. no. 2039/40, Q 7/8).

105. Metropolitan Museum of Art: hunting knife: inv. no. 04.3.158; maces: inv. nos. 04.3.39 and 14.25.1319.

106. Mann, 'Lost Armoury II', 58–59 (no. 84 of Libro di Aquila) and 68–69 (no. 177 of Libro di Aquila).

107. ASF, MdP 522, fol. 199r (MAP doc. ID 19527).

108. ASF, GM 43, p. 8. An inventory of the previous year, ASF, GM 45 (1560), fols. 54v–55r, however, records three boar spears, each with two small arquebuses.

109. ASC, Orsini, ser. II, 1205; KHM, Hofjagd- und Rüstkammer, inv. no. A 1189.

110. ASP, Casa e corte farnesiana, b. 52 (Inventari), fasc. 1; 1545 inventory of Ottavio Farnese.

111. ASF, GM 30, fol. ciij. Also recorded in 1560: GM 44, fol. 54r.

112. ASF, GM 30, fol. ciij.

113. ASF, GM 44, fol. 54r; GM 45 (1560), fol. 53r.

114. ASF, GM 43, p. 23.

115. The display did not give an inventory number when I visited in 2024.

116. Mallett, 'Condottieri and Captains', esp. 82 on gunpowder weapons and chivalry. On nobility and military service more generally, see Donati, *L'idea di nobiltà in Italia*; and Donati, 'Profession of Arms and the Nobility in Spanish Italy'.

117. Fletcher, *Beauty and Terror*, 112, 249.

118. Goldthwaite, *Wealth and the Demand for Art*, 41–42.

119. Satia, *Empire of Guns*, 9–10.

Chapter Five

1. Vannucci, *Jan Van Der Straet*, cat. no. 5, p. 85.

2. For a recent synthesis of the literature on this topic, see Pessina, *L'organizzazione militare*, 183–84.

3. Bayley, *War and Society*, 3.

4. Bayley, *War and Society*, ch. 5, 219–315, esp. 236–38.

5. Mallett and Hale, *Military Organization*, 79.

6. See 'Documenti per servire alla storia della milizia', 383, for palazzo supplies, and 386 for the 10 per cent and the possibility of all-firearms companies. For the process of issuing arms, see ASF, Nove conservatori di ordinanza e milizia, Distribuzione di armi 1. For context, see Andrea Guidi, *Un segretario militante*.

7. 'Documenti per servire alla storia della milizia', 383; Tlusty, *Martial Ethic*, 2.

8. For a wide-ranging survey of late medieval urban martial culture, see Schmid and Jaquet, *Cities in Arms* (forthcoming).

9. Magli, 'Walking the Streets'.

10. Boone, 'Towns, States and Martial Culture(s)'.

11. Moitzi, 'Vienna's Military Organization'.

12. Roelofson, 'Fighting for the Town'.

13. Ait, 'Per il controllo militare delle terre della Chiesa'; Najemy, 'Machiavelli and Cesare Borgia'; Byatt, 'Prince, Villain, Fortune's Fool'; Bayley, *War and Society*, 246.

14. 'Documenti per servire alla storia della milizia', 392, 429, 477.

15. *Il tiro al segno in Italia*, 78–79.

16. Bowd, *Renaissance Mass Murder*, 84–85.

17. *Il tiro al segno in Italia*, 76.

18. *Il tiro al segno in Italia*, app., p. li, doc. xlviii.

19. This was the War of the League of Cambrai; see Mallett and Shaw, *Italian Wars*, chs. 3–4.

20. Sanuto, *I diarii*, 47:307.

21. Mallett and Hale, *Military Organization*, 353, relying on Sanudo's figure, estimate the size of the Brescian force in 1528 at about a thousand, substantially lower than the four thousand figure from the 1530 ducal register: ASBs, CPI, Registri ducali 1, fol. 70r. This suggests that the numbers of militiamen increased rapidly: this is not implausible in light of concerns at the time that the Papal States might solicit Imperial support to secure the return of disputed territory from Venice.

22. Pessina, *L'organizzazione militare*, 186, 195.

23. In November 1528: *Il tiro al segno in Italia*, app., pp. lii–liii, doc. xlix, and discussion, 77, citing the observation of Benedetto Varchi, official historian to Duke Cosimo I de' Medici, that out of three thousand militiamen, seventeen hundred had been arquebusiers.

24. ASF, Otto di Pratica del Principato 121. For context, see Bayley, *War and Society*, 284–315.

25. ASF, MdP 633, fol. 11r (MAP doc. ID 9481). The total excludes men in the cities of Florence and Pistoia, or away at war.

26. *Documenti inediti*, 585–86.

27. 'Armi distribuite e vendute alle comunità Stato Ecclesiastico nel 1550': ASR, SG 88. In this case, the communities were required to pay for the weapons issued from their own funds.

28. Davis, 'Up in Smoke'; Ruff, *Violence in Early Modern Europe*, 65. For the same phenomenon in the seventeenth century, see Rose, *Renaissance of Violence*, 75–76, 189.

29. Mallett and Shaw, *Italian Wars*, 208; Sherer, *Warriors for a Living*, 36, 40–45.

30. *Documenti per servire alla storia della milizia italiana*, 397, 401–2.

31. Horn, *Jan Cornelisz Vermeyen*, 2: pl. xxix. Ramusio, *Navigazioni*, 4:508, 511 (Guagnino's *Descrizione della Sarmazia Europea*), first published 1578.

32. In Italian, *non vadino a male* or *in sinistro*: 'Disporre per la proprietà degli arredi militari degli Archibusieri morti e regolarmente cassata 12.4.1533': ASBs, CPI, Registri ducali 1, fol. 212r–v.

33. ASBo, Tribunale del Torrone (1560–1602), b. 1134. For the investigation as a whole see fols. 28r–29v, 65r–67v, 264r–267r and 270r–330v. For the party incident see fols. 275v, 278r, 280v, 288v, 290r, 292r, 294v, 301r, 312v, 315v, 320r, 325v.

34. ASBo, Tribunale del Torrone (1560–1602), b. 1134, fol. 307r.

35. ASBo, Tribunale del Torrone (1560–1602), b. 1134, fol. 309r.

36. ASBo, Tribunale del Torrone (1560–1602), b. 1134, fols. 297r, 301r, 309r, 326r.

37. Mallett and Hale, *Military Organization*, 203.

38. For an overview, see Delle Luche, *Des amitiés ciblées*; the quotation is the title of the conclusion (p. 249). Crombie, *Archery and Crossbow Guilds*, 162, notes that for archers and crossbowmen similar contests were established in the Low Countries by 1330.

39. ASVe, Deputati sopra il palio del falconetto, dello schioppo e della balestra 1, book of the *schioppo* contest, fol. 3v.

40. ASVe, Deputati sopra il palio, fol. 2r.

41. ASVe, Deputati sopra il palio, fols. 9r–10r.

42. ASVe, Deputati sopra il palio, fol. 5r.

43. *Il tiro al segno in Italia*, 44.

44. ASVe, Deputati sopra il palio, fol. 5v.

45 Mallett and Shaw, *Italian Wars*, 209–11.

46. ASVe, Deputati sopra il palio, fol. 7r–7v. For a similar case in 1527, see ASVe, CCX, Notatorio, Registri 7, fol. 132r.

47. Machiavelli, *Prince*, 50; see also the treatise on hunting by Ippolito de' Medici: Archivio Apostolico Vaticano, Miscellanea, Arm. II 78, fols. 317r–333v.

48. Vannucci, *Jan van der Straet*, cat. nos. 325, 326, 336, 341, and, for the collected *Cacce*, no. 693. Further discussion is in the recent exhibition catalogue, *Giovanni Stradano*, nos. 27, 28, 36.

49. ASBo, Tribunale del Torrone (1560–1602), b. 520, fol. 221v. The full case is at fols. 218r–232v and 242r–246r.

50. ASBo, Tribunale del Torrone (1560–1602), b. 520, fols. 223r and 229r.

51. Rose, *Renaissance of Violence*, 89. For examples of the use of firearms in Genoese feuds, see Raggio, *Feuds and State Formation*.

52. Tryner, '"Rascal with His Firestick"'.

53. ASVe, CCX, Dispacci dei Rettori (Brescia), 21, no. 189.

54. ASVe, CCX, Dispacci dei Rettori (Brescia), 22, nos. 206–24.

55. ASBo, Tribunale del Torrone (1560–1602), b. 1002, fol. 217r.

56. On *banditi*, see C. Shaw, *Politics of Exile*, 68–69; Davis, 'Up in Smoke', 399–400.

57. Fletcher, *Black Prince*, 180.

58. Fletcher, *Black Prince*, 181.

59. ASVe, CCX, Dispacci dei Rettori (Brescia), 20, no. 63.

60. Mallett and Hale, *Military Organization*, 354.

61. De' Lancellotti, *Cronaca Modenese*, 6:287, cited in Madden, *Civil Blood*, ch. 4.

62. ASBo, Legato, Bandi 1, fol. 340v.

63. Mallett and Shaw, *Italian Wars*, 209–11.

64. ASF, GM 7, fol. 47r; GM 31, fol. 75r.

65. ASC, Fondo Orsini, ser. I, vol. 412/2, fasc. 77.

66. ASC, Fondo Orsini, ser. I, vol. 412/2, fasc. 65.

67. For further discussion of the limitations of these records, see Bartels, 'Dressed to Kill'.

68. ASS, Curia del Placito 270, 1052, fol. 96v. See 'Fucile [Rifle/pistol]' in Hothi et al., *Refashioning the Renaissance Database*.

69. Garzoni, *Piazza di tutte le professioni del mondo*, 795–98.

70. ASF, Magistrato dei pupilli, 2653, fol. 692r. See 'Pugnale [Dagger]' in Hothi et al., *Refashioning the Renaissance Database*.

71. ASMo, ASE, Cancelleria ducale, Camera, Amministrazione della casa, Armeria, b. 1, reg. 5.: 'Nota delle persone che dicono sono provviste delle armi a loro tassate', 22 December 1551.

72. Monluc, *Commentaires*, 2:55. For commentary, see Eisenbichler, *Sword and the Pen*, esp. ch. 3.

73. Tryner, '"Rascal with His Firestick"', 152–54.

74. ASBo, Legato, Bandi 4, fol. 146v.

75. Morin and Held, *Beretta*, 44, citing ASVe, Collegio, Relazioni, b. 37, fol. 4r–v, published in Tagliaferri, *Relazioni*, 41. See also Bartels, 'Masculinity, Arms and Armour', 99.

76. Schwoerer, *Gun Culture*, 125–41.

77. Boone, 'Towns, States, and Martial Culture(s)'.

78. Baker, *Fruit of Liberty*, 126.

79. Bartels, 'Masculinity, Arms and Armour', 222.

80. Cellini, *Autobiography*, 60; *Vita*, 72.

Chapter Six

1. ASVe, CCX, Dispacci dei Rettori (Brescia), 21, no. 155.

2. For an example, see the anonymous account in Sanuto, *I diarii*, 56:195–96. On the shifting meanings of liberty in the republican context, see Baker, *Fruit of Liberty*; and C. Shaw, 'Concepts of *Libertà*'.

3. ASBo, Legato, Licenze di portare armi (1552), fols. 47r, 44v, 42v and 38v respectively.

4. ASBo, Legato, Licenze di portare armi (1552), fol. 28r.

5. ASBo, Legato, Bandi 1, fols. 118v–119v.

6. ASF, Miscellanea Medicea 39, ins. 15. This treatise was probably written in the mid- to late 1570s; given that it relates to a papal initiative to restrict the use of wheellock weapons but is to be found in the Medici archives, one might speculate on an association with Ferdinando de' Medici (cardinal 1563–88). Davis dates it to the 1580s: 'Renaissance Goes Up in Smoke', 403. The reference to the king of Spain as His Catholic Majesty certainly places it after 1556, while the 'most recent wars' of the Venetians and the Turks are likely to be the War of Cyprus (1570–73). The style of writing, generally plain, without classical references and alluding to themes widely found in soldiers' literature of the period, such as the idea of national character in warfare, suggests the author more likely had a military background than a humanist literary one. A date in the mid- to late 1570s would tally with some wider European context: this was the period of the first substantial published debate on firearms in England.

7. The ban followed letters patent along the same lines to subjects in Styria the previous year: Morin, 'Origins of the Wheellock', 85–86. For context, see Tlusty, *Martial Ethic*, ch. 3.

8. For the Holy Roman Empire, see Brugh, *Gunpowder*, 66; for England, see Schwoerer, *Gun Culture*, ch. 3, 46–64.

9. *Documenti inediti*, 304–5, 307–9.

10. ASBo, Senato 1383-sec. XVIII, Diari (1549–1731), unità nr. 2, Provisionum, 1522–1528, fols. 12v–14r, cited in Tryner, '"Rascal with His Firestick"', 176.

11. For Ferrarese bans dated 1523–25, 1534 and 1546, see *Documenti inediti*, 309–10, 312–13, 321–22. For Bologna, see Tryner, '"Rascal with His Firestick"', 176–79; and for Modena, see Madden, *Civil Blood*, chs. 3–4.

12. Morin, 'Origins', 86, 88. For an example of the distribution of the decree to the Venetian territories, see ASBs, CPI, Registri ducali 1, fol. 195r. The reiteration is in ASBs, CPI, Registri ducali 2, fols. 188v–189r.

13. ASNa, Archivio Farnese 572 (Castro e Ronciglione), fasc. 3, fol. 411v.

14. *Regni Siciliae pragmaticarum sanctionum in duas divisarum partes commentariorum*, 1:368.

15. Gelli, *Gli archibugiari milanesi*, 67.

16. Quarenghi, *Tecno-cronografia*, 186; Bartels, 'Masculinity, Arms and Armour', 204–5.

17. ASBo, Legato, Bandi 1, fol. 121r.

18. Coniglio, *Il viceregno di Don Pietro di Toledo (1532–53)*, 1:7.

19. ASVe, CCX, Dispacci dei Rettori (Brescia), 21, no. 111.

20. ASF, GM 43, 91bis.

21. ASBo, Legato, Bandi 1, fol. 192v (1544).

22. Blastenbrei, 'Violence, Arms and Criminal Justice in Papal Rome', 70–71.

23. *Documenti inediti*, 343–44, 356.

24. Hunters also requested licences for crossbows, which evidently remained a popular alternative. Only one volume, for 1552, survives in the series ASBo, Legato, Licenze di portare armi, but examples of firearms licences are also to be found in ASBo, Legato, Expeditiones 26 (1546), fol. 15r; 45 (1556), fols. 2v, 14v. Volumes in this series do not map precisely to calendar years, and there are some lacunae. For a discussion of the corresponding series in Florence, see Bartels, 'Masculinity, Arms and Armour', 79–90.

25. ASBo, Legato, Bandi 1, fols. 242v–243r.

26. Dohrn-van Rossum, *History of the Hour*. For an example of a clockmaker supplying flasks of gunpowder, see Bartels, 'Masculinity, Arms and Armour', 255.

27. This is a German example, dated 1580–1600. The museum display does not give inventory numbers.

28. Schalkhausser, 'Peter Peck', 184.

29. ASBs, CPI, Registri ducali 2, fol. 161r–v.

30. ASBo, Legato, Bandi 1, fol. 191v.

31. ASBs, CPI, Registri ducali 2, fol. 189r.

32. Morin, 'Origins', 84–85.

33. *Documenti inediti*, 324. For examples in Bologna, see ASBo, Legato, Bandi 1, fols. 118v, 195r, 205v. Phrases like 'of whatever status, rank or condition' are widely used in these proclamations.

34. ASBs, CPI, Registri ducali 3, fol. 14v.

35. ASBs, CPI, Registri ducali 4, fol. 35r.

36. ASBs, CPI, Registri ducali 3, fol. 15r.

37. ASBs, CPI, Registri ducali 3, fol. 22r–v.

38. Tagliaferri, *Relazioni*, 40.

39. ASBs, CPI, Registri ducali 2, fols. 13v, 39r; 3, fol. 23r–v; 4, fol. 148r; 5, fol. 190r–v.

40. An extended version of this discussion appears in Fletcher, 'Venice, Brescia and Gardone Val Trompia' (forthcoming).

41. Morin, 'Origins', 88.

42. ASVe, CCX, Dispacci dei Rettori (Brescia), 20, no. 54.

43. ASF, MdP 3719, fol. 772r (MAP doc. ID 23877).

44. ASF, MdP 3718, unnumbered folio (MAP doc. ID 23879).

45. ASF, MdP 1850, fol. 529r (MAP doc. ID 20154).

46. Morin, 'Origins', 88.

47. See, for example, ASVe, CCX, Lettere 26, no. 61, a letter in May 1526 to Padua, re scholars carrying arms after lobbying from the ambassador of the archduke of Austria.

48. ASVe, CCX, Notatorio, Registri 8, fol. 149v.

49. ASBo, Legato, Bandi 1, fol. 195v.

50. ASBo, Legato, Bandi 1, fols. 192r, 195r, 395r.

51. For some prices of arquebuses in the 1550s, see ASR, SG 88 (Conti Straordinari, 1541–52), insert on the Siege of Mirandola (1551–52). Prices include 225 gold ducats for two hundred guns. Exchange rates fluctuated but there were about 7 lire to the ducat (and 20 soldi to the lira).

52. ASBo, Legato, Expeditiones 72, fol. 66v.

53. ASBo, Tribunale del Torrone, Registri 1560–1602, b. 1134, fol. 77r.

54. ASCBs, Archivio Gambara 609, *Registro delle spese di casa di Lucrezio e Nicolò Gambara*, fol. 18v.

55. Porcellaga, fol. 93r; ins. 1 (of same busta), fol. 138r.

56. See Fletcher, *Diplomacy*, ch. 7.

57. ASBs, CPI, Registri ducali 2, fol. 189r.

58. ASMo, ASE, Cancelleria ducale, Archivi militari 50, F6, unnumbered insert, July 1557.

59. ASNa, Archivio Farnese 572 (Castro e Ronciglione), fasc. 2, fol. 154v.

60. ASVe, CCX, Dispacci dei Rettori (Brescia), 22, nos. 66–69. For his *condotte*, see Damiani, 'Sforza Pallavicini'.

61. This had also been a matter of dispute in 1543: ASVe, CCX, Dispacci dei Rettori (Brescia), 20, no. 170.

62. ASBs, CPI, Registri ducali 4, fol. 118v.

63. ASVe, CCX, Dispacci dei Rettori (Brescia), 21, no. 167.

64. ASNa, Archivio Farnese 281 (Varie), fol. 99r.

65. ASNa, Archivio Farnese 573 (Castro e Ronciglione), fol. 115r. An earlier set of ordinances is in ASNa, Archivio Farnese 572 (Castro e Ronciglione), fols. 157r–162v.

66. ASNa, Archivio Farnese 573, fols. 109r–120v (quote at 111v).

67. ASNa, Archivio Farnese 573, fol. 112v.

68. ASNa, Archivio Farnese 573, fol. 113r.

69. ASBo, Legato, Bandi 1, fols. 340r–341v.

70. ASBs, CPI, Registri ducali 2, fol. 151r–v.

71. ASBs, CPI, Registri ducali 2, fol. 129r–v.

72. ASBs, CPI, Registri ducali 2, fol. 161r–v.

73. ASBs, CPI, Registri ducali 3, fol. 97v.

74. ASBs, CPI, Registri ducali 3, fol. 172r.

75. ASF, MdP 210, fol. 68r (MAP doc. ID 8471).

76. ASBo, Tribunale del Torrone, Registri 1560–1602, b. 520, fols. 54r–62r, 66r–v.

77. Cellini, *Autobiography*, 120, incorrectly implies this was a flintlock; *Vita*, 132.

78. ASCBs, Archivio Gambara, Registri, b. 610, *Conto della spesa fatta per il viaggio a Roma* (unpaginated insert). For background on this text, see Bettoni, *I Gambara*; and Amighetti, 'Ambizioni curiali'.

79. ASBo, Tribunale del Torrone, Registri 1560–1602, b. 1002, fol. 128v.

80. ASBo, Legato, Bandi 6, fol. 129r. For similar systems in Rome and Perugia, see Davis, 'Up in Smoke', 405–6. For Florence, see Bartels, 'Masculinity, Arms and Armour', 67–68; and Bartels, 'Dressed to Kill'.

81. *Documenti inediti*, 324.

82. ASF, Miscellanea Medicea 39, ins. 15, fols. 1v–2r.

83. Tagliaferri, *Relazioni*, 40.

84. Davis, 'Up in Smoke', 403.

85. Davis, 'Up in Smoke', 408–9.

86. ASBs, CPI, Registri ducali 6, fol. 162r–62v.

87. Davis, 'Up in Smoke', 405–7.

Chapter Seven

1. Fletcher, 'Mere Emulators', 22.

2. Schwoerer, *Gun Culture*, ch. 3.

3. Tlusty, *Martial Ethic*, 81–87.

4. Fletcher, *Black Prince*, 3–5, 195–96, 199–204; Rebecchini, *"Un altro Lorenzo"*, 132–35; Dall'Aglio, *Duke's Assassin*, 3–6 and 165–66.

5. Marshall, 'Robert Pakington'; on the circulation of *Acts and Monuments*, see the essay by David Loades, 'Early Reception'.

6. *Letters and Papers Foreign and Domestic of the Reign of Henry VIII*, 1: no. 1463 [3496].

7. Sicca, 'Consumption and Trade of Art', 169.

8. Sicca and Waldman, *Anglo-Florentine Renaissance*; Woolfson, *Padua and the Tudors*.

9. *Calendar of State Papers Venice*, 5: no. 308; Schwoerer, *Gun Culture*, 106.

10. National Archives, State Papers 1/139, fol. 19r (*Letters and Papers*, 13.2: no. 828).

11. *Letters and Papers*, 12.2: no. 617, p. 227; 18.1, no. 630, p. 370.

12. Mayer, 'Reginald Pole'. For background, see Mayer, *Reginald Pole*.

13. Porcellaga, fol. 169v.

14. Romani, *Le corti farnesiane*, 1:29. For additional discussion, see Bertelli, 'Il principe nuovo'; Quarenghi, *Tecno-cronografia*, 1:184.

15. Bertoncini, *La congiura*, 11.

16. Ricci, *Gli atti*, 45–46; see also 52.

17. Ricci, *Gli atti*, 50; also summarized on 53.

18. Ricci, *Gli atti*, 81.

19. Ricci, *Gli atti*, 82.

20. Ricci, *Gli atti*, 102.

21. Ricci, *Gli atti*, 101.

22. Ricci, *Gli atti*, 104.

23. Ricci, *Gli atti*, 107.

24. Ricci, *Gli atti*, 108.

25. Ricci, *Gli atti*, 112.

26. Medici, *Apology*, 30. For context, see Dall'Aglio, *Duke's Assassin*, 159–76.

27. Medici, *Apology*, 23.

28. Medici, *Apology*, 42.

29. Medici, *Apology*, 31.

30. Sutherland, *Princes, Politics and Religion*, 139.

31. G. B. Possevino, *Discorsi della vita*, 142–56.

32. *Vera relatione*, 42; see also Giussano, *Istoria*, 101; Bascapé, *De vita*, 72.

33. El Messiry and Eltahan, 'Enhancement of Silk Fabric'; Oleksiak, 'Monk Who Stopped Bullets with Silk'.

34. Giussano, *Istoria*, 100.

35. *Vera relatione*, 38.

36. Giussano, *Istoria*, 100.

37. Royal Collection, inv. no. RCIN 902400. See PatER–Catalogo del Patrimonio culturale dell'Emilia-Romagna, 'Attentato a San Carlo'.

38. Birrel, 'Diarey', 18. Scots spellings slightly Anglicized.

39. Anonymous *Diurnal*, 156.

40. Melville, *Memoirs*, 223.

41. Calderwood, *History of the Kirk*, 2:510–11.

42. On this murder, see Sutherland, *Princes, Politics and Religion*, 139–55.

43. Sutherland, 'Assassination', 279.

44. Sutherland, 'Assassination', 280, 286.

45. Knecht, *French Civil Wars*, 88.

46. Holt, *French Wars of Religion*, 62.

47. Paré, *Oeuvres complètes*, 3:724; translation from Wood, *King's Army*, 199.

48. Knecht, *French Civil Wars*, 81.

49. *Archives curieuses*, ser. 1, 5:168. Carroll, *Martyrs and Murderers*, 167. See also Sutherland, *Princes, Politics and Religion*, 139–55.

50. *Calendar of State Papers, Foreign*, 6: no. 354; Forbes, *Full View of Public Transactions*, 2:339.

51. Holt, *French Wars of Religion*, 82–83.

52. Jouanna, *Saint Bartholomew's Day*, 73–74; Carroll, *Martyrs and Murderers*, 192–93; Estebe, *Tocsin*, 107–8.

53. Goulart, *Mémoirs de l'estat de France*, 1:367.

54. Jouanna, *Saint Bartholomew's Day*, 73; Estebe, *Tocsin*, 108.

55. Kingdon, *Myths*, 15–16. For the account of Coligny's shooting in these pamphlets, see 28–29.

56. Davis, 'Up in Smoke', 405–6.

57. See Hale's introduction to Smythe, *Certain Discourses Military*, xli–lvi.

58. Ramusio, *Navigazioni* 4:421, 556. These observations are from Alessandro Guagnino, 'La descrizione della Sarmazia europea'.

59. Jardine, *Awful End*, 54–55. Ridley, *William the Silent*, 91–101, also provides a narrative of the assassination attempts.

60. Jardine, *Awful End*, 64–72.

61. Jardine, *Awful End*, 56–7, 149. Some historians suggest that he had more than one firearm; see Ridley, *William the Silent*, 101, n. 10.

62. ASBo, Legato, Bandi 4, fols. 134r–135v.

63. It was not until 2014 that an international arms trade treaty came into force, and its coverage remains limited; see United Nations Office for Disarmament Affairs, 'Arms Trade'.

Chapter Eight

1. Hale, *Artists and Warfare*, 95.

2. Guerriero, *Atlante*.

3. Capitoline Museums (Rome), inv. no. PC 40.

4. Hale, *Artists and Warfare*, esp. chs. 4 and 5.

5. For the curators' note, see Galleria Doria Pamphilj, 'Tapestry of Alexander the Great'. On the Pastrana tapestries and Grenier, see *Invention of Glory*, 9.

6. Anon., *Contested Landing of Alexander the Great in Scythia*, miniature, ?c. 1480, Bibliothèque Nationale de France, MS fr. 6440, f. 173. In Hale, *Artists and Warfare*, 44, fig. 64.

7. Devisse and Mollat, *Image of the Black*, 2.2:171, and fig. 151, 332 n. 82.

8. Devisse and Mollat, *Image of the Black*, 2.2:168.

9. Münzer, 'Itinerarium', 87, cited in Saunders, *Social History of Black Slaves and Freedmen*, 83. I am grateful to Jenny Bulstrode for drawing this reference to my attention.

10. Devisse and Mollat, *Image of the Black*, 2.2:171.

11. For a survey of these representations, see Suckale-Redlefsen, *Black St Maurice*.

12. Devisse and Mollat, *Image of the Black*, 2.2:70–72.

13. For a wide-ranging discussion of premodern histories of race and visual culture, with references, see Ndiaye and Markey, *Seeing Race before Race*.

14. Hale, *Artists and Warfare*, 46, fig. 68.

15. Rijksmuseum, inv. no. SK-C-402.

16. On the later civic guard paintings, see Adams, *Public Faces and Private Identities*, ch. 5, 211–58.

17. On the transition, see Knevel, *Burgers in het geweer*.

18. Adams, *Public Faces and Private Identities*, 215. See also Falomir, *El Retrato*, 127.

19. Amsterdam Museum, inv. no. SA 7278.

20. On the 'flight to the land' and its much-debated implications, see Franco Cazzola, 'Il "ritorno alla terra"'.

21. The Titian original is lost but a copy is in the Metropolitan Museum of Art, inv. no. 27.56. The Dossi portrait is in the Galleria Estense (Modena).

22. ASMo, ASE, Camera, Amministrazione della casa, Armeria, b. 1, regg. 3 and 4.

23. Giovio, *Vita di Alfonso da Este*, 16, and the poem by Francesco Sperulo, *Villa Julia Medica*, in *Raphael in Early Modern Sources*, i, 431; Taylor, *Art of War*, 91–92. Discussed in Fletcher, *Beauty and the Terror*, 141–42.

24. The quotation is given in the catalogue entry at the Victoria and Albert Museum, inv. no. 29404:34. See also Strauss, *Illustrated Bartsch*, 10:421, cited in Stiber, Eusman and Albro, 'Triumphal Arch'. Other contributors to the work included Albrecht Altdorfer, Hans Springinklee, Hans Beck and Hans Scheufelein.

25. *Battle of Wenzenbach*, c. 1515, Staatliche Graphische Sammlung (Munich); Ingersoll, 'Battle of Wenzenbach', 184–85. For further context to this series, see Cuneo, 'Images of Warfare'.

26. Falomir, *El Retrato*, 388; the description is from the 1558 posthumous inventory of Mary of Hungary.

27. Musée de l'Armée (Paris), inv. no. 815 PO. I am grateful to Hans Busio for drawing my attention to this connection, on which see Schalkhausser, 'Peter Peck', 194.

28. Uffizi Gallery, Inventory 1890, no. 1439.

29. Panofsky, *Problems in Titian*, 84–87; Pedrocco, *Titian*, 206; Mancini, *Tiziano*, 37–41; Wethey, *Paintings of Titian*, 35–36.

30. Falomir, *El Retrato*, 388–89 (for an English translation, see the Museo del Prado website: https://www.museodelprado.es/en). The gun is not mentioned in either Humfrey, *Titian*, 158, or Da Campo, *Art of Power*, 220–21.

31. Museo del Prado (Madrid), inv. no. P001184.

32. Vannucci, *Jan Van Der Straet*, cat. nos. 325 (deer), 326 (ibex), 336 and 341 (geese).

33. Museo di Roma, inventory number not supplied on label.

34. Link made in interpretive material, Capitoline Museums, on display December 2023. For Bril's imagery of firearms in hunting scenes see for example Cappelletti, *Paul Bril*, 161, cat. no. 145.

35. Frick Collection (New York), inv. no. 2021.2.147.

36. For discussion, see Mulcahy, 'Enea Vico's Proposed Triumphs'; and Sani, 'War on a Plate'. The engraving is in the Royal Collection, inv. no. RCIN 721016, with further copies in the Metropolitan Museum of Art and British Museum.

37. On the Renaissance study and collecting, see D. Thornton, *Scholar in His Study*, 99–125.

38. Sansovino, *Venetia*, 137–39. For context, see D. Thornton, *Scholar in His Study*, 114.

39. National Gallery, inv. no. NG292. I am grateful to Wouter Wagemakers for drawing this image to my attention.

40. Alte Pinakothek (Munich), inv. no. 5352.

41. Sternbersky Palace, inv. no. VO 1193.

42. Staatsgalerie (Stuttgart); KHM; Copenhagen Art Museum—all in Hale, *Artists and Warfare*, 243–44 (fig. 311), 256, 257–58 (fig. 334). I have found a claim that a gun features in the 1520s frescoes at the Cappella della Crocifissione at Sacro Monte, however the gun is not apparent in the photographs I have consulted. See Royal Armouries, 'Matchlock Military Musket'.

43. Stracke, 'Palazzo Bellomo Altarpiece of St. Barbara'.

44. Gatta, *Per grazia ricevuta*, cat. nos. 58 and 63. I am grateful to Mary Laven for drawing my attention to these examples.

45. A. Possevino, *Soldato christiano*, 52–57. Possevino's work can be viewed online at Gallica, the digital library of the Bibliothèque Nationale de France (https://gallica.bnf.fr). For context, see Brunelli, *Soldati del Papa*, 11–17.

46. Haag, *Bruegel the Master*, 248–51; Roberts-Jones and Roberts-Jones, *Bruegel*, 128–34.

47. Haag, *Bruegel the Master*, 191.

48. Pieter Bruegel the Elder, *Resurrection*, Walker Art Gallery, inv. no. 1177; *Conversion of St Paul*, KHM, Gemäldegalerie, inv. no. 3690.

49. Haag, *Bruegel the Master*, 250.

50. Kaschek, Müller and Buskirk, *Pieter Bruegel the Elder and Religion*, 3–4.

51. Schmitz-von Ledebur, 'Emperor Charles V Captures Tunis'.

52. Horn, *Jan Cornelisz Vermeyen*, 1:26–27. For the images, see 2: pls. A65 and A66.

53. Soykut, *Italian Perceptions*, 65–67; Giovio, *Notable Men and Women*, 23.

54. Horn, *Jan Cornelisz Vermeyen*, 2: pls. A127 and A128.

55. Muccini, *Salone del Cinquecento*, 54.

56. Robertson, *Il Gran Cardinale*, 97, and, on the series as a whole, 95–102.

57. Strunck, 'Barbarous and Noble Enemy', has the most detailed discussion of this theme. Broader collections include Stagno and Llopis, *Lepanto and Beyond*; and Fuchs and Weissbourd, *Representing Imperial Rivalry*.

58. Ramusio, *Navigazioni*, 3:517, 4:152–53.

59. Ramusio, *Navigazioni*, 3:372–73.

60. ASR, SG, 87, fols. 50r, 54r, 55r.

61. Williams, 'Mediterranean Conflict', 53.

62. Ágoston, *Guns for the Sultan*, 23–24.

63. *Süleymanname*, pls. 12–14.

64. Ágoston, *Guns for the Sultan*, 3.

65. See Rota, 'Il Vero ritratto di Zarra et di Sebenico'.

66. Strunck, 'Barbarous and Noble Enemy', 222.

67. Museo del Prado (Madrid), inv. no. P000431.

68. KHM, Gemäldegalerie, inv. no. 8270.

69. *Battle of Lepanto*, 40–41.

70. *Battle of Lepanto*, 66–67, 72–73.

71. *Battle of Lepanto*, 234–35.

72. *Battle of Lepanto*, 262–63.

73. *Battle of Lepanto*, 264–65.

74. *Battle of Lepanto*, 266–67.

75. *Battle of Lepanto*, 270–71.

76. Musée cantonal des Beaux-Arts de Lausanne, inv. no. 729.

77. Rijksmuseum, inv. no. RP-P-OB-78.785-238.

78. Tom Hamilton, 'Procession of the League'. See *Procession de la Ligue sur l'île de la Cité*, Musée Carnavalet (Paris), inv. no. P622. Hamilton convincingly questions the museum's attribution to François Bunel.

79. Tanis and Horst, *Images of Discord*, 26.

80. Tanis and Horst, *Images of Discord*, 42, cat. no. 3. The Antichrist's entourage is described in German as his household and in French as his soldiers.

81. Sold at auction, 2001; see Christie's, 'Follower of Anthonis Mor van Dashorst'.

82. In the collection of the Royal Palace Amsterdam, inv. no. SC/0129.

83. On Tresham, see English Heritage, 'History of Rushton Triangular Lodge'. I am grateful to Peter Wilson for drawing my attention to this image. The Dudley portrait is at Parham House; the Drury portrait at Boughton House. Goldring, *Robert Dudley*, 152 and fig. 128. On Stanley: Jardine, *Awful End*, 91–92. The British Museum and National Portrait Gallery have reidentified their portraits of Stanley as showing Robert Dudley.

84. Goldring, *Robert Dudley*, ch. 5, 123–64, esp. 123, 148.

85. Goldring, *Robert Dudley*, 146.

86. Jardine, *Awful End*, 90.

87. Vanucci, *Jan Van Der Straet*, cat. no. 505, now in Teylers Museum (Haarlem); cat. no. 506 is its preparatory drawing, now in the Cooper Hewitt Museum, New York. The full series is in the Bibliothèque Nationale (Paris) cat. No. 704.

88. Tate Gallery, inv. no. T03028 (details from catalogue entry).

89. See Rijksmuseum, inv. no. SK-C-378.

90. On the lost paintings, see Blackwood, 'Meta Incognita'.

91. Oviedo, in Ramusio, *Navigazioni*, 5:576.

92. Small, 'Transformer and Influencer', 40.

93. Schwoerer, *Gun Culture*, 116–17.

94. Rijksmuseum, inv. no. SK-C-375.

Chapter Nine

1. Ramusio, *Navigazioni*, 2:688.

2. Cook, *Hundred Years' War*, 73.

3. Chase, *Firearms*, 138.

4. Chase, *Firearms*, 141, 145; Irwin, 'Gunpowder and Firearms', 119.

5. Irwin, 'Gunpowder and Firearms', 139.

6. See the essays in Raudzens, *Technology, Disease and Colonial Conquests*, for contextual discussion.

7. Kang, 'Korean Snap Matchlock'.

8. Silverman, *Thundersticks*; Satia, *Empire of Guns*; Green, *Fistful of Shells*.

9. Jacob and Visoni-Alonzo, *Military Revolution*.

10. Cook, *Hundred Years' War*, 1–2.

11. Smith, *Warfare and Diplomacy*, 81. Cook's account of Morocco's access to gunpowder technology suggests that Smith's proposal that guns' value 'lay as much in their novelty and noise as in their lethal quality' may need to be revisited.

12. Cook, *Hundred Years' War*, 121, citing Pulgar, *Crónica*, 2:127–28.

13. Cook, *Hundred Years' War*, 109.

14. Pereira, *Esmeraldo de situ orbis*, 20, cited in Cook, *Hundred Years' War*, 137.

15. Africanus, *Description de l'Afrique*, 2:424 (also in Ramusio, *Navigazioni*, 1:352). For context, see Cook, *Hundred Years' War*, 171, who, however, partially mistranslates the source.

16. *Sources inedités*, 1.2:223, cited in Cook, *Hundred Years' War*, 265; see 261 for context.

17. On the impact of the work, see Small, 'Transformer and Influencer'.

18. Ramusio, *Navigazioni*, 1:634–35.

19. Ramusio, *Navigazioni*, 1:632.

20. De Varthema, in Ramusio *Navigazioni*, 1:812, 817, 879.

21. Barbosa, in Ramusio, *Navigazioni*, 2:539 (editor's note).

22. Barbosa, in Ramusio, *Navigazioni*, 2:580–81 (elephants), 580 (Brahmins), 612 (Lingayats). The Lingayats are a Hindu sect.

23. For examples of gun salutes, see Ramusio, *Navigazioni*, 1:693 and 749.

24. Pigafetta, in Ramusio, *Navigazioni*, 2:935.

25. Pigafetta, in Ramusio, *Navigazioni*, 2:930.

26. Africanus, in Ramusio, *Navigazioni*, 1:388; Smith, *Warfare and Diplomacy*, 81.

27. Alvarez, in Ramusio, *Navigazioni*, 2:87.

28. Alvarez, in Ramusio, *Navigazioni*, 2:223, 226, 228–29, 258–59.

29. Ramusio, *Navigazioni*, 2:376.

30. Alvarez, in Ramusio, *Navigazioni*, 2:359.

31. Pory, 'Briefe relation', 984.

32. Jacob and Visoni-Alonzo, *Military Revolution*, 38.

33. Pigafetta, in Ramusio, *Navigazioni*, 2:907–9.

34. De' Fedrici, in Ramusio, *Navigazioni*, 6:1062–63; Cesare de' Fedrici was there in 1569 although the account was not published until 1587 (pp. 1015, 1070).

35. Ramusio, *Navigazioni*, 1:629.

36. Giovanni da Empoli, in Ramusio, *Navigazioni*, 1:745.

37. Giovanni da Empoli, in Ramusio, *Navigazioni*, 1:889.

38. De Marees, *Description and Historical Account*, intro., xiii, xv.

39. De Marees, *Description and Historical Account*, 91. For the persistence of these narratives of fear into the seventeenth century, see *German Sources for West African History*, 63, 67, 96.

40. De Marees, *Description and Historical Account*, 92; n. 8 explains that a Gotelingh was a cast twelve-pounder gun.

41. De Marees, *Description and Historical Account*, 220. The combination of 'slaves' and 'subjects' here introduces some ambiguity. The editor suggests this may refer to the Asafo companies of Elmina.

42. Chase, *Firearms*, 110; *German Sources for West African History*, 91.

43. *German Sources for West African History*, 95.

44. Radburn, 'British Gunpowder Industry'.

45. Townsend, *Fifth Sun*, 125.

46. *Italian Reports*, 59.

47. Martire, in Ramusio, *Navigazioni*, 5:61.

48. Martire, in Ramusio, *Navigazioni*, 5:109.

49. Martire, in Ramusio, *Navigazioni*, 5:136.

50. The first part of the *Natural and General History* was published in 1535, with the first book of the second part following in 1557; the rest was not published until the nineteenth century; the *Summary* was published in 1526. Ramusio, *Navigazioni*, 5:236–37 and 253–54 (Sommario); 435–36 (Istoria).

51. For overviews of the conquests, see Cervantes, *Conquistadores*; and Townsend, *Fifth Sun*.

52. Cortés, in Ramusio, *Navigazioni*, 6:136, 139.

53. Cortés, in Ramusio, *Navigazioni*, 6:148.

54. Cortés, in Ramusio, *Navigazioni*, 6:161.

55. Jacob and Visoni-Alonzo, *Military Revolution*, 36.

56. *Italian Reports*, 27–28.

57. Ramusio, *Navigazioni*, 1:675.

58. Ramusio, *Navigazioni*, 1:626.

59. Giovanni da Empoli, in Ramusio, *Navigazioni*, 1:744.

60. Martire, in Ramusio, *Navigazioni*, 5:32, a point reiterated on p. 37.

61. Martire, in Ramusio, *Navigazioni*, 5:83.

62. Oviedo, in Ramusio, *Navigazioni*, 5:456, 523, 796.

63. Ramusio, *Navigazioni*, 6:866 (letter from Oviedo to Cardinal Bembo).

64. Martire, in Ramusio, *Navigazioni*, 5:111, 145.

65. Martire, in Ramusio, *Navigazioni*, 5:100.

66. Oviedo, in Ramusio, *Navigazioni*, 5:761–62.

67. Ramusio, *Navigazioni*, 6:866.

68. Montaigne, *Essays* (1991), 1030; see discussion in Grafton, *New Worlds, Ancient Texts*, 155.

69. In addition to those cited already, see the relation of an anonymous Spanish captain involved in the conquest of Peru in Ramusio, *Navigazioni*, 6:688–89.

70. Silverman, *Thundersticks*, 23.

71. De Ulloa, in Ramusio, *Navigazioni*, 6:538 (crossbow bolts); Cortès, in Ramusio, *Navigazioni*, 6:176, 178 (powder).

72. Oviedo, in Ramusio, *Navigazioni*, 5:789.

73. Oviedo, in Ramusio, *Navigazioni*, 5:858.

74. D'Alvarado, in Ramusio, *Navigazioni*, 6:314; see n. for measurements.

75. Ramusio, *Navigazioni*, 6:345 (attribution), 347 (metals).

76. Oviedo, in Ramusio, *Navigazioni*, 5:557.

77. Oviedo, in Ramusio, *Navigazioni*, 5:806, 808.

78. Alvarez, in Ramusio, *Navigazioni*, 2:227.

79. Cortés, in Ramusio, *Navigazioni*, 6:290–91.

80. Oviedo, in Ramusio, *Navigazioni*, 5:940.

81. See Catalogo generale dei Beni Culturali, 'Motivi decorativi a grottesche', for a catalogue of the images.

82. Heikamp, *Mexico and the Medici*, 19–22; Markey, *Imagining the Americas*, ch. 7, 93–117.

83. For the chronology, Peterson and Terraciano, *Florentine Codex*, 3, and for the collaborators, 6 and 13. I have followed the terminology used in this volume.

84. Rao, 'Reception of the Florentine Codex', 37–39.

85. Terraciano, 'Reading between the Lines of Book 12', 45.

86. Terraciano, 'Reading between the Lines of Book 12', 46; Escalante Gonzalbo, 'Art of War', 67–71.

87. Sahagún, *Florentine Codex*, 15–16. A reference on p. 41 to the 'great lombard guns which went resting on wooden wheels' makes it clear these were not handguns.

88. Sahagún, *Florentine Codex*, 19; Lockhart, *We People Here*, 80, gives a more informal English translation.

89. Sahagún, *Florentine Codex*, 62, 72, 85–86.

90. Sahagún, *Florentine Codex*, 121.

91. Terraciano, 'Reading between the Lines of Book 12', 67.

92. See images 48 and 49 of Book 12 of the *Florentine Codex* in the Library of Congress World Digital Library (http://www.wdl.org/); subsequent citations are to this edition.

93. Book 12, image 93.

94. Book 12, images 113, 125, 114.

95. Book 12, image 81.

96. The Digital Benin website (www.digitalbenin.org) has catalogued fourteen figures of Portuguese soldiers in museum collections internationally, primarily produced in the seventeenth and eighteenth centuries but often featuring earlier styles of armour.

97. For the debate as it has played out in Britain, see Hicks, *Brutish Museums*; B. Phillips, *Loot*; and Lundén, 'Distorting History'. A critique of the Eurocentric nature of this discussion is in Moiloa, *Reclaiming Restitution*.

98. Ezra, *Royal Art of Benin*; Barley, *Art of Benin*.

99. Donahue-Wallace, *Art and Architecture of Viceregal Latin America*, ch. 5; Bailey, *Andean Hybrid Baroque*, 312–13.

Conclusion

1 Just before the First World War, Fenner Brockway's play on this theme, *The Devil's Business*, was banned in London. See Brockway, *Bloody Traffic*; and, for context, Fletcher, 'Worst Business'.

2 Small Arms Survey, 'Global Firearms Holdings'.

BIBLIOGRAPHY

Primary Sources

MANUSCRIPT

Archivio Apostolico Vaticano

Miscellanea, Arm. II

Archivio di Stato di Bologna

Legato, Bandi
Legato, Expeditiones
Legato, Licenze di portare armi
Senato 1383-sec. XVIII, Diari (1549–1731)
Tribunale del Torrone, Registri (1560–1602)

Archivio di Stato di Brescia

Archivio Martinengo dalle Palle
Archivio Notarile
Cancelleria prefettizia inferiore, Registri ducali

Archivio di Stato di Firenze

Guardaroba Medicea
Mediceo del Principato
Miscellanea Medicea
Nove conservatori di ordinanza e milizia, distribuzione di armi
Nove conservatori di ordinanza e milizia, Entrata e uscita / Debitori e creditori
Otto di Pratica del Principato

Archivio di Stato di Modena

Archivio Segreto Estense, Cancelleria ducale, Archivi militari
Archivio Segreto Estense, Cancelleria ducale, Camera, Amministrazione della casa, Armeria

Archivio di Stato di Napoli

Archivio Farnese, Parma

Archivio di Stato di Parma

Casa e corte farnesiana

Archivio di Stato di Roma

Archivio Notarile
Soldatesche e Galere

Archivio di Stato di Venezia

Capi del Consiglio dei Dieci, Dispacci dei Rettori
Capi del Consiglio dei Dieci, Notatorio, Registri
Deputati sopra il palio
Patroni e provveditori all'Arsenale

Archivio Storico Capitolino, Rome

Fondo Orsini, ser. I and II.

Archivio Storico Civico di Brescia

Archivio Gambara

National Archives, London

SP 1/139 State Papers, 1

Museum Collections

Alte Pinakothek, Munich, Germany
Amsterdam Museum, Netherlands
Apostolic Palace, Vatican City
Art Institute of Chicago, US
Bayerisches Nationalmuseum, Munich, Germany
Bodleian Libraries, Oxford, UK
Boughton House, Northamptonshire, UK
Brescia Museums, Italy
British Museum, London, UK
Capitoline Museums, Rome, Italy
Copenhagen Art Museum, Denmark
Frick Collection, New York, US
Galleria Estense, Modena, Italy
Konopiště Castle, Prague, Czechia
Kunsthistorisches Museum, Vienna, Austria
Livrustkammaren, Stockholm, Sweden
Mary Rose Museum, Portsmouth, UK
Metropolitan Museum of Art, New York, US
Musée Cantonal des Beaux-Arts de Lausanne
Musée Carnavalet, Paris, France
Museo Correr, Venice, Italy

Musée de l'Armée, Paris, France
Museo Civico Luigi Marzoli, Brescia, Italy
Museo del Prado, Madrid, Spain
Museu Militar, Lisbon, Portugal
National Gallery, London, UK
National Gallery of Ireland, Dublin
Nationalmuseum, Stockholm, Sweden
Newberry Library, Chicago, US
Osterreichische Nationalbiblioteck, Vienna, Austria
Palazzo del Principe, Genoa, Italy
Palazzo Ducale, Venice, Italy
Palazzo Vecchio, Florence, Italy
Parham House, Sussex, UK
Real Armeria, Madrid, Spain
Rijksmuseum, Amsterdam, Netherlands
Royal Armouries, Leeds, UK
Royal Collection, UK
Royal Palace Amsterdam, Netherlands
Rushton Triangular Lodge, UK
Staatliche Graphisches Sammlung, Munich, Germany
Staatsgalerie, Stuttgart, Germany
Sternbersky Palace, Prague, Czechia
Tate Gallery, London, UK
Uffizi Gallery, Florence, Italy
Veneranda Biblioteca e Pinacoteca Ambrosiana, Milan
Victoria and Albert Museum, London, UK
Walker Art Gallery, Liverpool, UK
Wallace Collection, London, UK

Printed Primary Sources

Accademia della Crusca. *Vocabolario degli Accademici della Crusca*. Giovanni Alberti, 1612.

Africanus, Leo (al-Hasan ibn Muhammad al-Wazzan). *Description de l'Afrique*. Translated by A. Épaulard. 2 vols. Librarie d'Amérique et d'Orient Adrien-Maisonneuve, 1956.

Archives curieuses de l'histoire de France depuis Louis XI jusqu'à Louis XVIII. Edited by L. Cimber and F. Danjou. Beauvais, 1835.

Ariosto, Ludovico. *Orlando Furioso*. Edited by Douglas B. Killings. Translated by William Stewart Rose, 1995. https://atlas.cs.brown.edu/data/gutenberg/6/1/615/615.txt.

Ariosto, Ludovico. *Orlando Furioso secondo l'edizione del 1532*. Edited by Ottavio Morali. Pirotta, 1818.

Ascham, Roger. *English Works*. Edited by W. A. Wright. Cambridge University Press, 1904.

Barret, Robert. *The Theorike and Practike of Moderne Warres*. William Ponsonby, 1598.

Bascapé, C. *De vita et rebus gestis Caroli S. R. E. Cardinalis tituli S. Praxedis*. Davidis Sartorii, 1592.

The Battle of Lepanto. Edited and translated by Elizabeth R. Wright, Sarah Spence, and Andrew Lemons. Harvard University Press, 2014.

Biringuccio, Vanoccio. *Pirotechnia: I dieci libri della Pirotechnia*. Curzio Troiano Navo, 1558.

Birrel, Robert. 'The Diarey of Robert Birrel'. In *Fragments of Scottish History*, edited by J. G. Dalyell. Archibald Constable, 1798.

Calderwood, David. *The History of the Kirk of Scotland*. 8 vols. Wodrow Society, 1842–49.

Calendar of State Papers, Foreign Series, of the Reign of Elizabeth. 23 vols. HMSO, 1863–1950.

Calendar of State Papers Relating to English Affairs in the Archives of Venice. 38 vols. HMSO, 1864–1967.

Cardano, Girolamo. *The Book of My Life*. Translated by Jean Stoner. New York Review Books, 2002.

Castiglione, Baldassarre. *The Courtier*. Translated by George Bull. Penguin, 1967.

Cellini, Benvenuto. *Autobiography*. Translated by George Bull. Penguin, 1998.

Cellini, Benvenuto. *Vita di Benvenuto Cellini*. Edited by Orazio Bacci. Sansoni, 1901.

Commynes, Philippe de. *The Memoirs of Philip de Commines*. Edited and translated by Andrew R. Scobie. 2 vols. Bell, 1906.

De' Lancellotti, Tommasino de' Bianchi detto. *Cronaca Modenese*. Pietro Fiaccadori, 1862.

De Marees, Pieter. *Description and Historical Account of the Gold Kingdom of Guinea (1602)*. Edited and translated by Albert van Dantzig and Adam Jones. Oxford University Press, 1987.

Descriptio Urbis: The Roman Census of 1527. Edited by Egmont Lee. Bulzoni, 1985.

A Diurnal of Remarkable Occurrents in Scotland, AD MDXIII–AD MDLXXV. [Bannatyne Club], 1833.

Documenti inediti per la storia delle armi da fuoco italiane. Edited by Angelo Angelucci. Cassone, 1869; labelled vol. 1, pt. 1; no further parts published.

'Documenti per servire alla storia della milizia italiana dal XIII secolo al XVI raccolti negli archivi di Toscana'. Edited by Giuseppe Canestrini, special edition of *Archivio Storico Italiano*, ser. 1, 15 (1851).

Du Bellay, Joachim. *The Regrets*. Translated by David R. Slavitt. Northwestern University Press, 2004.

Erasmus of Rotterdam. *Collected Works*. 89 vols. Toronto University Press, 1974–.

Erasmus of Rotterdam. *The Complaint of Peace*. Open Court Publishing, 1917.

Fourquevaux, Raimond. *Instructions for the Warres*. Translated by Paule Iue. Thomas Man and Tobie Cooke, 1589.

A Full View of the Public Transactions in the Reign of Q. Elizabeth. Edited by Patrick Forbes. Vol. 2. Bettenham, 1741.

Fronsperger, Leonhard. *Kriegsbuch*. 3 vols. [Johannes Schmidt], 1573.

Gambara, Veronica. *Complete Poems*. Edited and translated by Molly M. Martin and Paola Ugolini. Centre for Reformation and Renaissance Studies, 2014.

Garimberto, Girolamo. *Il Capitano Generale*. Giordano Ziletti, 1556.

Garzoni, Tommaso. *Piazza di tutte le professioni del mondo*. Meghetti, 1605.

German Sources for West African History. Edited and translated by Adam Jones. Franz Steiner, 1983.

Gilino, Coradino. *The 'De Morbo quem Gallico nuncupant' (1497) of Coradinus Gilinus*. Edited and translated by Cyril C. Barnard. Brill, 1930.

Giovio, Paolo. *La vita di Alfonso da Este duca di Ferrara*. Translated by Giovambattista Gelli. Venice, 1597.

Giovio, Paolo. *Notable Men and Women of Our Times*. Edited and translated by Kenneth Gouwens. Harvard University Press, 2013.

Giussano, G. P. *Istoria della vita, virtù, morte e miracoli di Carlo Borromeo*. Giacomo Sarzina, 1615.

Gli atti del procedimento in morte di Pier Luigi Farnese: Un'istruttoria non chiusa. Edited by Aldo G. Ricci. Banca di Piacenza, 2007.

Goulart, Simon. *Mémoirs de l'estat de France, sous Charles neufiesme*. 3 vols. Heinrich Wolf, 1577.

Guicciardini, Francesco. *Carteggi di Francesco Guicciardini*. Edited by Roberto Palmarocchi. Istituto Storico Italiano per l'Eta Moderna e Contemporanea, 1951.

Guicciardini, Francesco. *The History of Italy*. Translated by Austin Parke Goddard. 3rd ed., 10 vols. London, 1763.

Hetoum. *A lytell cronycle: Richard Pynson's translation (c. 1520) of La fleur des histoires de la terre d'Orient (c. 1307)*. Edited by Glenn Burger. Toronto University Press, 1988.

Interiano, Giorgio. 'Vita et Sito de Zychi di Giorgio Interiano: Trascrizione e commento dell'editio princeps del 1502'. Edited by Francesco Crifò and Wolfgang Schweickard. *Zeitschrift für romanische Philologie* 130 (2014): 160–78.

The Inventory of King Henry VIII. Edited by David Starkey. 2 vols. Society of Antiquaries, 1998.

Italian Reports on America, 1493–1522: Accounts by Contemporary Observers. Edited by Geoffrey Symcox and Luciano Formisano. Translated by Theodore J. Cachey Jr. and John C. McLucas. Brepols, 2002.

Leonardo da Vinci. *Notebooks*. Selected by Irma Richter. Edited by Thereza Wells. Oxford University Press, 2008.

Les Sources inedités de l'histoire du Maroc. 1st series. Edited by Henry de Castries. Archives et Bibliothèques de France, 1909.

Letters and Papers, Foreign and Domestic, of the Reign of Henry VIII. Edited by J. S. Brewer, J. Gairdner and R. H. Brodie. 22 vols. HMSO, 1862–1932.

Lorenzo de' Medici at Home: The Inventory of the Palazzo Medici in 1492. Edited and translated by Richard Stapleford. Pennsylvania State University Press, 2013.

Machiavelli, Niccolò. *The Art of War*. Edited and translated by Christopher Lynch. Chicago University Press, 2003.

Machiavelli, Niccolò. *The Arte of Warre*. Translated by Peter Withorne. W. Williamson, 1573.

Machiavelli, Niccolò. *The Prince*. Edited by Peter Bondanella. Translated by Peter Bondanella and Mark Musa. Oxford University Press, 1984.

Medici, Lorenzino de'. *Apology for a Murder*. Translated by Andrew Brown. Hesperus Classics, 2004.

Melville, James. *Memoirs of His Own Life*. George Brookman, 1833.

Monluc, Blaise de. *Commentaires et lettres de Blaise de Monluc, Maréchal de France*. Edited by Alphonse de Rable. 5 vols. Renouard, 1864–72.

Montaigne, Michel de. *The Essayes of Michael Lord of Montaigne*. Edited by H. Morley. Translated by John Florio. Routledge, 1885.

Montaigne, Michel de. *The Essays of Michel de Montaigne*. Edited and translated by M. A. Screech. Allen Lane, 1991.

Mora, Domenico. *Il Soldato di M. Domenico Mora, bolognese, gentilhuomo grisone et cavalliere academico Stordito*. Gabriel Giolito de Ferrari, 1570.

Münzer, Hieronymus. 'Itinerarium'. *Revue Hispanique* 48 (1920): 1–179.

Nunziature di Venezia. Edited by Franco Gaeta and Aldo Stella. Istituto storico italiano per l'età moderna e contemporanea. 1958–

Paré, Ambroise. *Oeuvres complètes*. Edited by J. F. Malgaigne. Vol. 3. J. B. Baillière, 1841.

Pereira, Duarte Pacheco. *Esmeraldo de situ orbis*. Edited by George H. T. Kimble. London, 1937.

Peterson, Jeanette Favrot, and Kevin Terraciano, eds. *The Florentine Codex: An Encyclopedia of the Nahua World in Sixteenth-century Mexico*. University of Texas Press, 2019.

Pius II [Aeneas Sylvius Piccolomini]. *Secret Memoirs of a Renaissance Pope. The Commentaries of Aeneas Sylvius Piccolomini, Pius II*. Edited by Leona C. Gabel. Translated by Florence Alden Gragg. London, 1988.

Pory, John. 'A briefe relation concerning the dominions, revenues, forces, and maner of government of sundry the greatest princes either inhabiting within the bounds of Africa, or at least possessing some parts thereof.' In Leo Africanus, *The History and Description of Africa, vol. 3*, edited by Robert Brown, 973–1000. Cambridge University Press, 2010.

Possevino, Antonio. *Il soldato christiano*. Dorici, 1569.

Possevino, Giovan Battista. *Discorsi della vita, et attioni di Carlo Borromeo*. Tornieri, 1591.

Pulgar, Fernando del. *Crónica de los Reyes Católicos*. Edited by Juan de Mata Carriazo. 2 vols. Espasa-Colpe, 1943.

Ramusio, Giovanni Battista. *Navigazioni e viaggi*. 6 vols. Einaudi, 1978–c. 1983.

Raphael in Early Modern Sources, 1483–1602. Edited by John Shearman. Yale University Press, 2003.

Regni Siciliae pragmaticarum sanctionum in duas divisarum partes commentariorum. Edited by Mario Muta. Francisco Ciotto, 1622.

Rich, Barnabe. *A Right Exelent and Pleasaunt Dialogue, Betwene Mercury and an English Soldier*. 1574.

Il Sacco di Brescia: testimonianze, cronache, diari, atti del processo e memorie storiche della "presa memoranda et crudele" della città nel 1512. Edited by Vasco Frati. Comune di Brescia, 1989.

Sahagún, Fray Bernardino de. *Florentine Codex: General History of the Things of New Spain*. Translated by Arthur J. O. Anderson and Charles E. Dibble. University of Utah Press, 2012.

Sansovino, Francesco. *Venetia, città nobilissima et singolare*. Iacomo Sansovino, 1581.

Sanudo [Sanuto], Marin. *I diarii*. Edited by Rinaldo Fulin et al. 59 vols. Forni Editore, 1969–70.

Sanudo [Sanuto], Marin. *Itinerario per la Terraferma veneziana*. Edited by Gian Maria Varanini. Viella, 2014.

Smythe, Sir John. *Certain Discourses Military*. Edited by J. R. Hale. Cornell University Press, 1964.

Spandouyn, Theodore [Spandugino]. *Petit Traicté de l'origine des Turcqz*. Edited by C. Schéfer. Ernest Laroux, 1896.

State Papers Published Under the Authority of Her Majesty's Commission. Volume VII. King Henry the Eighth Part V: Continued. Record Commission, 1849.

Süleymanname: The Illustrated History of Süleyman the Magnificent. Edited by Esin Atil. National Gallery of Art, Washington, DC; Harry N. Abrams, 1986.

Taboni Facchini, Clara, and Anna Tanghetti, eds. *Gli annali del Voltolino*. In *Bovegno di Valle Trompia: Fonti per una storia*, 9–23. Grafo, 1985.

Tagliaferri, A., ed. *Relazioni dei Rettori Veneti in Terraferma. XI, Podestaria e Capitanato di Brescia*. Giuffrè, 1978.

Tartaglia, Nicolo. *Three Bookes of Colloquies Concerning the Arte of Shooting in Great and Small Peeces of Artillerie*. Translated by Cyprian Lucar. John Harrison, 1588.

Tartalea [Tartaglia], Nicolo. *Quesiti et inventioni diverse*. Venturino Ruffinelli, 1546.

Van Wesenbecke, Jacques. *Mémoires de Jacques de Wesenbeke*. Edited by C. Rahlenbeck. M. Weissenbruch, 1859.

Vera relatione del successo dell'archibuggiata tirata a S. Carlo Borromeo arcivescovo di Milano. Edited by Giulia Bologna. La Vita Felice, 1995.

We People Here: Nahuatl Accounts of the Conquest of Mexico. Edited and translated by James Lockhart. University of California Press, 1993.

Withorne, Peter. *Certaine Wayes for the Ordering of Souldiours in Battelray*. W. Williamson, 1573.

Secondary Literature

Adams, Ann Jensen. *Public Faces and Private Identities in Seventeenth-century Amsterdam: Portraiture and the Production of Community*. Cambridge University Press, 2009.

Ágoston, Gábor. *Guns for the Sultan: Military Power and the Weapons Industry in the Ottoman Empire*. Cambridge University Press, 2005.

Ágoston, Gábor. 'War-winning Weapons? On the Decisiveness of Ottoman Firearms from the Siege of Constantinople (1453) to the Battle of Mohács (1526)'. *Journal of Turkish Studies* 39 (2013): 129–43.

Ait, Ivana. 'Per il controllo militare delle terre della Chiesa: l'*Hermandad* di Alessandro VI, organizzazione e finanziamento'. In *Alessandro VI e lo stato della Chiesa: atti del convegno, Perugia, 13–15 marzo 2000*, edited by Carla Frova and Maria Grazia Nico Ottaviani, 37–77. Ministero per i beni e le attività culturali, Direzione generale per gli archivi, 2003.

Ajmar-Wollheim, Marta, and Flora Dennis, eds. *At Home in Renaissance Italy: Art and Life in the Italian House, 1400–1600*. V&A Publications, 2006.

Amighetti, Paolo Maria. 'Ambizioni curiali e carriera diplomatica di Francesco Gambara tra Roma, Venezia e l'Impero (1576–1630)'. In *I Gambara e Brescia nell'Italia del tardo Rinascimento: Diplomazia, mecenatismo, cultura e consumi*, edited by Barbara Bettoni, 69–89. FrancoAngeli, 2019.

Andrade, Tonio. *The Gunpowder Age: China, Military Innovation and the Rise of the West in World History*. Princeton University Press, 2016.

Angelucci, Angelo. *Il tiro al segno in Italia dalla sua origine sino ai nostri giorni*. Baglione, 1865.

Ansani, Fabrizio. 'The Life of a Renaissance Gunmaker: Bonaccorso Ghiberti and the Development of Florentine Artillery in the Late Fifteenth Century'. *Technology and Culture* 58 (2017): 749–89.

Antologia gardonese. Apollonio, 1969.

Arfaioli, Maurizio. *The Black Bands of Giovanni: Infantry and Diplomacy during the Italian Wars (1526–1528)*. Edizioni Plus / Pisa University Press, 2005.

Armi e cultura nel bresciano, 1420–1870. Ateneo di Brescia, 1981.

'Armoured Soldiers Firing Match-Lock Arquebus, Late 15th Century (Hand Coloured Woodcut)'. Private Collection, Bridgeman Images. Accessed 1 August 2025. https://www.bridgemanimages.com/en-US/german-school/armoured-soldiers-firing-match-lock-arquebus-late-15th-century-hand-coloured-woodcut/hand-coloured-woodcut/asset/386843.

Arnold, Thomas F. 'Fortifications and the Military Revolution: The Gonzaga Experience, 1530–1630'. In *The Military Revolution Debate: Readings on the Military Transformation of Early Modern Europe*, edited by Clifford J. Rogers, 201–26. Westview Press, 1995.

Arrivo, Georgia. 'Una dinastia al femminile. Per uno sguardo diverso sulla storia politico-istituzionale'. In *Carte di donne: per un censimento regionale della scrittura delle donne dal XVI al XX secolo*, edited by Anna Scattigno and Alessandra Contini, 49–58. Edizioni di storia e letteratura, 2007.

Arrizabalaga, Jon, John Henderson and Roger French. *The Great Pox: The French Disease in Renaissance Europe*. Yale University Press, 1997.

Auslander, Leora, and Tara Zahra. *Objects of War: The Material Culture of Conflict and Displacement*. Cornell University Press, 2018.

Bailey, Gauvin. *The Andean Hybrid Baroque: Convergent Cultures in the Churches of Colonial Peru*. University of Notre Dame Press, 2010.

Baker, Nicholas Scott. *The Fruit of Liberty: Political Culture in the Florentine Renaissance, 1480–1550*. Harvard University Press, 2013.

Barbiroli, Bruno. *Repertorio storico degli archibugiari italiani dal XIV al XX secolo*. CLUEB, 2012.

Barley, Nigel. *The Art of Benin*. British Museum, 2010.

Bartels, Victoria. 'Dressed to Kill. Arms, Armour and Protective Attire in Renaissance Men's Middle- and Lower-class Dress'. In *Refashioning the Renaissance: Everyday Dress in Europe, 1500–1650*, edited by Paola Hohti, 170–89. Manchester University Press, 2025.

Bartels, Victoria. 'Masculinity, Arms and Armour, and the Culture of Warfare in Sixteenth-century Florence'. Unpublished PhD dissertation, University of Cambridge, 2019.

Bayley, C. C. *War and Society in Renaissance Florence: The* De Militia *of Leonardo Bruni*. Toronto University Press, 1961.

Belfanti, Carlo Marco. 'A Chain of Skills: The Production Cycle of Firearms Manufacture in the Brescia Area from the Sixteenth to the Eighteenth Centuries'. In *Guilds, Markets and Work Regulations in Italy, 16th–19th Centuries*, edited by Alberto Guenzi, Paola Massa and Fausto Piola Caselli, 266–83. Ashgate, 1998.

Bercusson, Sarah. 'Gift-giving, Consumption and the Female Court in Sixteenth-century Italy'. Unpublished PhD thesis, Queen Mary University of London, 2009.

Bertelli, Sergio. 'Il principe nuovo'. In *La congiura farnesiana dopo 460 anni: una rivolta contro lo Stato nuovo*, edited by Marco Bertoncini, 43–61. Banca di Piacenza, 2008.

Bertolotti, Antonio. 'Le arti minori alla corte di Mantova nei secoli XV, XVI e XVII'. *Archivio Storico Lombardo*, 2nd ser, 5 (1888): 491–590.

Bertoncini, Marco, ed. *La congiura farnesiana dopo 460 anni: una rivolta contro lo Stato nuovo*. Banca di Piacenza, 2008.

Bettoni, Barbara, ed. *I Gambara e Brescia nell'Italia del tardo Rinascimento: Diplomazia, mecenatismo, cultura e consumi*. FrancoAngeli, 2019.

Blackwood, Nicole. 'Meta Incognita: Some Hypotheses on Cornelis Ketel's Lost English and Inuit Portraits'. *Netherlands Yearbook for the History of Art* 66 (2016): 28–53.

Blair, Claude. 'Further Notes on the Origin of the Wheellock'. *Arms and Armour Annual* 1 (1973): 28–47.

Blastenbrei, Peter. 'I romani tra violenza e giustizia nel tardo Cinquecento'. *Roma Moderna Contemporanea* 1 (1997): 67–79.

Blastenbrei, Peter. 'Violence, Arms and Criminal Justice in Papal Rome'. *Renaissance Studies* 20 (2006): 68–87.

Blockmans, Wim, André Holenstein and Jon Mathieu. *Empowering Interactions: Political Cultures and the Emergence of the State in Europe, 1300–1900*. Routledge, 2009.

Bolognini, Pierantonio. 'La produzione e l'organizzazione del lavoro nelle fucine gardonesi'. In *Armi antiche a Gardone*, edited by Cesare Calamandrei, 14–24. Fondazione Negri, 2008.

Boone, Marc. 'Towns, States and Martial Culture(s): Some Considerations Inspired by the History of the Urban Militias of the Former Low Countries'. In *Cities in Arms: Urban Martial Culture in Late Medieval and Early Modern Europe*, edited by Regula Schmid and Daniel Jaquet. Boydell and Brewer, forthcoming.

Bossini, F., and M. Galeri, eds. *Valtrompia nella storia*. Compagnia della Stampa, 2007.

Bowd, Stephen D. *Renaissance Mass Murder: Civilians and Soldiers during the Italian Wars*. Oxford University Press, 2018.

Bowd, Stephen D. *Venice's Most Loyal City: Civic Identity in Renaissance Brescia*. Harvard University Press, 2010.

Bowd, Stephen D., Sarah Cockram and John Gagné, eds. *Shadow Agents of Renaissance War: Suffering, Supporting, and Supplying Conflict in Italy and Beyond*. Amsterdam University Press, 2023.

Brigden, Susan. 'Henry VIII and the Crusade against England'. In *Henry VIII and the Court*, edited by Thomas Betteridge and Suzannah Lipscomb, 215–34. Routledge, 2013.

Brockway, Fenner. *The Bloody Traffic*. Gollancz, 1933.

Brooker, Robert E. 'What Can Be Learned from the Landeszeughaus Wheellock Collection in Graz, Austria'. *American Society of Arms Collectors Bulletin* 95 (2007): 9–18. https://americansocietyofarmscollectors.org/wp-content/uploads/2019/06/2007-B95-What-Can-be-Learned-from-the-Landeszeugh.pdf.

Brugh, Patrick. *Gunpowder, Masculinity, and Warfare in German Texts, 1400–1700*. University of Rochester Press, 2019.

Brunelli, Giampiero. 'Camillo Orsini'. In *Dizionario biografico degli italiani*. Treccani, 2013. https://www.treccani.it/enciclopedia/camillo-orsini_(Dizionario-Biografico).

Brunelli, Giampiero. *Soldati del Papa: politica militare e nobiltà nello Stato della Chiesa: 1560–1644*. Carocci, 2003.

Buchanan, Iain. 'The "Battle of Pavia" and the Tapestry Collection of Don Carlos: New Documentation'. *The Burlington Magazine* 144 (2002): 345–51.

Butters, H. C., and Gabriele Neher, eds. *Warfare and Politics: Cities and Government in Renaissance Tuscany and Venice*. Amsterdam University Press, 2020.

Byatt, Lucinda. 'Prince, Villain, Fortune's Fool: Is Cesare Borgia's Reputation beyond Repair?' In *The Borgia Family: Rumor and Representation*, edited by Jennifer Mara DeSilva, 193–222. Routledge, 2020.

Cappelletti, Franceesca. *Paul Bril e la pittura di paesaggio a Roma 1580–1630*. Ugo Bozzi, 2005–2006.

Capra, Carlo. 'The Italian States in the Early Modern Period'. In *The Rise of the Fiscal State in Europe, c.1200–1815*, edited by Richard Bonney, 413–42. Oxford University Press, 1999.

Cardarelli, François. *Encyclopaedia of Scientific Units, Weights and Measures: Their SI Equivalences and Origins*. Springer, 2003.

Carpegna, Nolfo di. *Antiche armi dal sec IX al XVIII gia Collezione Odescalchi: catalogo*. De Luca, 1969.

Carpegna, Nolfo di. *Brescian Firearms*. De Luca, 1997.

Carr, Christopher. *Kalashnikov Cultures*. Praeger, 2008.

Carroll, Stuart. *Enmity and Violence in Early Modern Europe*. Cambridge University Press, 2023.

Carroll, Stuart. *Martyrs and Murderers: The Guise Family and the Making of Europe*. Oxford University Press, 2009.

Caselli, Fausto Piola. 'The Formation of Fiscal States in Italy: The Papal States'. In *The Rise of Fiscal States: A Global History, 1500–1914*, edited by Bartolomé Yun-Casalilla and Patrick K. O'Brien with Francisco Comín Comín, 285–319. Cambridge University Press, 2012.

Catalogo generale dei Beni Culturali. 'Motivi decorativi a grottesche'. Accessed 9 February 2025. https://catalogo.beniculturali.it/detail/HistoricOrArtisticProperty/0900281221-1.

Cazzola, Franco. 'Il "ritorno alla terra"'. In *Storia della società italiana*, vol. 10, *Il tramonto del Rinascimento*, edited by Giovanni Cherubini et al., 103–68. Teti, 1987.

Cervantes, Fernando. *Conquistadores: A New History*. Allen Lane, 2020.

Chase, Kenneth. *Firearms: A Global History to 1700*. Cambridge University Press, 2003.

Chilton, Paul. 'Humanism and War in Rabelais and Montaigne'. In *War, Literature and the Arts in Sixteenth-century Europe*, edited by Margaret Shewring and J. R. Mulryne, 119–43. Macmillan, 1989.

Christie's. 'Follower of Anthonis Mor van Dashorst'. Accessed 23 July 2024. https://www.christies.com/lot/lot-follower-of-anthonis-mor-van-dashorst-portrait-2030115/.

Cohn, Samuel K., Jr., and Fabrizio Ricciardelli, eds. *The Culture of Violence in Renaissance Italy*. Le Lettere, 2012.

Cominazzi, Marco. *Cenni sulla fabbrica d'armi in Gardone di Valtrompia*. Sentinella, 1845.

Coniglio, Giuseppe. *Il viceregno di Don Pietro di Toledo (1532–53)*. 2 vols. Giannini, 1984.

Cook, Weston F. *The Hundred Years War for Morocco: Gunpowder and the Military Revolution in the Early Modern Muslim World*. Westview Press, 1994.

Covini, Maria Nadia. 'Political and Military Bonds in the Italian State System, Thirteenth to Sixteenth Centuries'. In *War and Competition between States*, edited by Philippe Contamine, 9–36. Oxford University Press, 2000.

Cressy, David. *Saltpeter: The Mother of Gunpowder*. Oxford University Press, 2013.

Crombie, Laura. *Archery and Crossbow Guilds in Medieval Flanders, 1300–1500*. Boydell and Brewer, 2018.

Cuneo, Pia F. 'Images of Warfare as Political Legitimization: Jörg Breu the Elder's Rondels for Maximilian I's Hunting Lodge at Lermos (ca. 1516)'. In *Artful Armies, Beautiful Battles: Art and Warfare in Early Modern Europe*, edited by Pia Cuneo, 87–105. Brill, 2002.

Da Campo, Soler, ed. *The Art of Power: Royal Armor and Portraits from Imperial Spain*. Translated by Jenny Dodman. SEACEX, Patrimonio Nacional, TF Editores, 2009.

Dall'Aglio, Stefano. *The Duke's Assassin: The Exile and Death of Lorenzino de' Medici*. Yale University Press, 2015.

Damiani, Roberto. 'Sforza Pallavicini'. *Condottieri di Ventura*, 27 November 2012. https://condottieridiventura.it/sforza-pallavicini-di-fiorenzuola-d/.

Davies, Jonathan, ed. *Aspects of Violence in Renaissance Europe*. Routledge, 2013.

Davis, Robert C. 'The Renaissance Goes Up in Smoke'. In *The Renaissance World*, edited by John Jeffries Martin, 435–48. Routledge, 2008.

Dean, Trevor, 'Eight Varieties of Homicide'. In *Murder in Renaissance Italy*, edited by Trevor Dean and K. J. P. Lowe, 83–105. Cambridge University Press, 2017.

Delle Luche, J.-D. *Des amities ciblées: concours de tir et diplomatie urbaine dans le Saint-Empire, XVe–XVIe siècle*. Brepols, 2021.

Delle Luche, J.-D. 'Pouvoir, fabricants et sociétés de tir face à la diffusion des canon rayés dans le Saint-Empire (2e moitié du XVIe siècle)'. *Francia* 46 (2019): 147–66.

Del Torre, Giuseppe. *Venezia e la Terraferma dopo la guerra di Cambrai: fiscalità e amministrazione (1515–1530)*. F. Angeli, 1986.

Devisse, Jean, and Michel Mollat, eds. *The Image of the Black in Western Art*, vol. 2, *From the Early Christian Era to the "Age of Discovery"*. Cambridge, MA: Harvard University Press, 2010.

De Vries, Kelly. 'Military Surgical Practice and the Advent of Gunpowder Weaponry'. *Canadian Bulletin of Medical History* 7 (1990): 131–46.

Dohrn-van Rossum, Gerhard. *History of the Hour: Clocks and Modern Temporal Orders*. University of Chicago Press, 1996.

Donahue-Wallace, Kelly. *Art and Architecture of Viceregal Latin America, 1521–1821*. University of New Mexico Press, 2008.

Donati, Claudio. *L'idea di nobiltà in Italia, secoli XIV–XVIII*. Laterza, 1995.

Donati, Claudio. 'The Profession of Arms and the Nobility in Spanish Italy: Some Considerations'. In *Spain in Italy: Politics, Society, and Religion*, edited by Thomas J. Dandelet and John A. Marino, 299–324. Brill, 2007.

Eisenbichler, Konrad. *The Sword and the Pen: Women, Politics and Poetry in Sixteenth-century Siena*. University of Notre Dame Press, 2012.

Eisenstein, Elizabeth L. *The Printing Press as an Agent of Change*. Cambridge University Press, 1980 (online edition, 2013).

Elias, Norbert. *The Civilizing Process, Sociogenetic and Psychogenetic Investigations*. Revised edition. Edited by Eric Dunning, Johan Goudsblom and Stephen Mennell. Translated by Edmund Jephcott. Blackwell, 2000.

El Messiry, Magdi, and Eman Eltahan. 'Enhancement of Silk Fabric Knife-Stabbing Resistance for Soft Body Armor'. *Journal of Industrial Textiles* 54 (2024): 1–29.

Eltis, David. *The Military Revolution in Sixteenth-century Europe*. Tauris, 1998.

English Heritage. 'History of Rushton Triangular Lodge'. Accessed 29 January 2025. https://www.english-heritage.org.uk/visit/places/rushton-triangular-lodge/history/.

Escalante Gonzalbo, Pablo. 'The Art of War, the Working Class, and Snowfall: Reflections on the Assimilation of Western Aesthetics'. In *The Florentine Codex: An Encyclopedia of the Nahua World in Sixteenth-century Mexico*, edited by Jeanette Favrot Peterson and Kevin Terraciano, 63–74. University of Texas Press, 2019.

Estebe, Janine. *Tocsin pour une Massacre: La Saison des Saint-Barthélemy*. Editions du Centurion, 1968.

Ezra, Kate. *Royal Art of Benin: The Perls Collection*. Metropolitan Museum of Art, 1992.

Falomir, Miguel. *El Retrato del Renacimiento*. Museo Nacional del Prado, 2008.

Fantoni, Marcello. *La Corte del Granduca: Forma e simboli del potere mediceo fra Cinque e Seicento*. Bulzoni, 1994.

Fantoni, Marcello. 'Un Rinascimento a metà. Le corti italiane nella storiografia anglo-americana'. *Cheiron* 17 (1997): 1–31.

Ferraro, Joanne M. *Family and Public Life in Brescia, 1580–1650: The Foundations of Power in the Venetian State*. Cambridge University Press, 1993.

Ferraro, Joanne M. 'Feudal-Patrician Investments in the Bresciano and the Politics of the *Estimo*, 1426–1641'. *Studi Veneziani* 7 (1983): 31–57.

Fletcher, Catherine. 'Agents of Arms Supply in Early Modern Italy: Constructing the Military State'. In *Shadow Agents of Renaissance War*, edited by Stephen Bowd, Sarah Cockram and John Gagné, 201–26. Amsterdam University Press, 2022.

Fletcher, Catherine. *The Beauty and the Terror: An Alternative History of the Italian Renaissance*. Bodley Head, 2020.

Fletcher, Catherine. *The Black Prince of Florence: The Spectacular Life and Treacherous World of Alessandro de' Medici*. Bodley Head, 2016.

Fletcher, Catherine. *Diplomacy in Renaissance Rome: The Rise of the Resident Ambassador*. Cambridge University Press, 2015.

Fletcher, Catherine. '"Furnished with Gentlemen": The Ambassador's House in Sixteenth-century Italy'. *Renaissance Studies* 24 (2010): 518–35.

Fletcher, Catherine. 'Mere Emulators of Italy: The Spanish in Italian Diplomatic Discourse, 1492–1550'. In *The Spanish Presence in Sixteenth-century Italy: Images of Iberia*, edited by Piers Baker-Bates and Miles Pattenden, 11–28. Ashgate, 2014.

Fletcher, Catherine. 'Venice, Brescia and Gardone Val Trompia: Martial Culture in a Subject City and Its Hinterland'. In *Cities in Arms: Urban Martial Culture in Late Medieval and Early Modern Europe*, edited by Regula Schmid and Daniel Jaquet. Boydell and Brewer, forthcoming.

Fletcher, Catherine. 'The Worst Business in the World? The Emotional Historiography of the Arms Industry'. In *The Business of Emotions in Modern History*, edited by Andrew Popp and Mandy Cooper, 177–94. Bloomsbury, 2023.

Foley, Vernard, Steven Rowley, David F. Cassidy and F. Charles Logan. 'Leonardo, the Wheel Lock, and the Milling Process'. *Technology and Culture* 24 (1983): 399–427.

Foucault, Michel. *Discipline and Punish: The Birth of the Prison*. Translated by Alan Sheridan. Pantheon Books, 1977.

Franci, Riccardo. *Le armi da fuoco a ruota al Museo Stibbert: Wheellock Firearms at the Stibbert Museum*. Stibbert Museum, 2016.

Fuchs, Barbara, and Emily Weissbourd, eds. *Representing Imperial Rivalry in the Early Modern Mediterranean*. Toronto University Press, 2015.

Fynn-Paul, Jeff, ed. *War, Entrepreneurs, and the State in Europe and the Mediterranean, 1300–1800*. Brill, 2014.

Gaibi, Agostino. *Armi da fuoco italiane dal Medioevo al Risorgimento*. Bramante, 1978.

Galleria Doria Pamphilj. 'The Tapestry of Alexander the Great'. Accessed 1 August 2025. https://www.doriapamphilj.it/en/genoa/the-art/the-tapestries/.

Gatta, Annalisa, et al., eds. *Per Grazia Ricevuta: Gli ex-voto di San Nicola a Tolentino*. Biblioteca Egidiana, 2005.

Gelli, Jacopo. *Gli archibugiari milanesi*. Hoepli, 1905.

Giovanni Stradano: Le più strane e belle invenzioni del mondo. Edited by Alessandra Baroni. Edizioni Polistampa, 2023.

Goertz, Hans-Jürgen. *The Anabaptists*. Routledge, 1996.

Goldring, Elizabeth. *Robert Dudley, Earl of Leicester and the World of Elizabethan Art: Painting and Patronage at the Court of Elizabeth I*. Yale University Press, 2014.

Goldthwaite, Richard. *Wealth and the Demand for Art in Italy, 1300–1600*. Johns Hopkins University Press, 1993.

Grafton, Anthony, with April Shelford and Nancy Siraisi. *New Worlds, Ancient Texts: The Power of Tradition and the Shock of Discovery*. Harvard University Press, 1995.

Green, Toby. *A Fistful of Shells: West Africa from the Rise of the Slave Trade to the Age of Revolution*. Penguin, 2020.

Greener, W. W. *The Gun and Its Development*. 3rd ed. Cassell, 1881.

Guerriero, Simone, ed. *Atlante delle xilografie italiane del Rinascimento*. Fondazione Giorgio Cini—Istituto di Storia dell'Arte. Accessed 27 August 2025. https://archivi.cini.it/storiaarte/archive/IT-SDA-GUI001-000038/atlante-xilografie-italiane-del-rinascimento.html.Guerrini, P. *La congregazione dei padri della pace*. Scuola Tipografica Opera Pavoniana, 1933.

Guidi, Andrea. *Un segretario militante: politica, diplomazia e armi nel Cancelliere Machiavelli*. Il Mulino, 2009.

Guilmartin, John F. 'The Military Revolution: Origins and First Tests Abroad'. In *The Military Revolution Debate: Readings on the Military Transformation of Early Modern Europe*, edited by Clifford J. Rogers, 299–333. Westview Press, 1995.

Haag, Sabine, ed. *Bruegel the Master*. Thames and Hudson, 2018.

Hale, John. *Artists and Warfare in the Renaissance*. Yale University Press, 1990.

Hale, John. 'Gunpowder and the Renaissance'. In *From the Renaissance to the Counter-Reformation: Essays in Honor of Garrett Mattingly*, edited by C. H. Carter, 113–44. Random House, 1965.

Hall, Alfred. *Ballistics in the Seventeenth Century: A Study in the Relations of Science and War with Reference Principally to England*. Cambridge University Press, 1952.

Hall, Bert S. *Weapons and Warfare in Renaissance Europe: Gunpowder, Technology and Tactics*. Johns Hopkins University Press, 1997.

Hamilton, Tom. 'The Procession of the League: Remembering the Wars of Religion in Visual and Literary Satire'. *French History* 30 (2016): 1–30.

Hayward, J. F. *The Art of the Gunmaker*. 2 vols. Barrie and Rockliff, 1962.

Hayward, Maria. *Dress at the Court of Henry VIII*. Maney, 2007.

Heikamp, Detlef. *Mexico and the Medici*. Edam, 1972.

Hicks, Dan. *The Brutish Museums: The Benin Bronzes, Colonial Violence and Restitution*. Pluto Press, 2020.

Hohti, Paula, et al. *Refashioning the Renaissance Database: Popular Groups and the Material and Cultural Significance of Clothing in Europe 1550–1650*. 2017–22. https://refashioningdata.eu/home.

Holt, Mack P. *The French Wars of Religion, 1562–1629*. Cambridge University Press, 2005.

Horn, Hendrik J. *Jan Cornelisz Vermeyen: Painter of Charles V and His Conquest of Tunis*. 2 vols. Davaco, 1989.

Humfrey, Peter. *Titian*. Phaidon, 2007.

Ingersoll, Catharine. 'The Battle of Wenzenbach as a Pictorial Subject at the Habsburg and Wittelsbach Courts'. In *Imagery and Ingenuity in Early Modern Europe: Essays in Honor of Jeffrey Chipps Smith*, edited by Catharine Ingersoll, Alisa McCussker and Jessica Weiss, 181–90. Brepols, 2018.

Innes, Joanna. 'The Regulation of Charity and the Rise of the State'. In *The Routledge History of Poverty, c.1450–1800*, edited by David Hitchcock and Julia McClure, 3–20. Routledge, 2021.

The Invention of Glory: Afonso V and the Pastrana Tapestries. Fundación Carlos de Amberes / Ediciones El Viso, 2011.

Irwin, R. 'Gunpowder and Firearms in the Mamluk Sultanate Reconsidered'. In *The Mamluks in Egyptian and Syrian Politics and Society*, edited by M. Winter and A. Levanoni, 117–39. Brill, 2004.

Jacob, Frank, and Gilmar Visoni-Alonzo. *The Military Revolution in Early Modern Europe: A Revision*. Palgrave, 2016.

Jardine, Lisa. *The Awful End of Prince William the Silent: The First Assassination of a Head of State with a Handgun*. Harper Collins, 2005.

Jouanna, Arlette. *The Saint Bartholomew's Day Massacre: The Mysteries of a Crime of State*. Translated by Joseph Bergin. Manchester University Press, 2015.

Jütte, Robert. 'Poor Relief and Social Discipline in Sixteenth-century Europe'. *European Studies Review* 11 (1981): 25–52.

Jütte, Robert. *Poverty and Deviance in Early Modern Europe*. Cambridge University Press, 1994.

Kang, Hyeok Hweon. 'The Korean Snap Matchlock: A Global Microhistory'. *American Society of Arms Collectors Bulletin* 124 (2022): 30–41.

Karonen, Petri, and Marko Hakanen. *Personal Agency at the Swedish Age of Greatness, 1560–1720*. Finnish Literature Society, 2017.

Kaschek, Bertram, Jürgen Müller, and Jessica Buskirk. *Pieter Bruegel the Elder and Religion*. Brill, 2018.

Kea, R. A. 'Firearms and Warfare on the Gold and Slave Coasts from the Sixteenth to the Nineteenth Centuries'. *Journal of African History* 12 (1971): 185–213.

Kingdon, Robert M. *Myths about the St Bartholomew's Day Massacres 1572–1576*. Harvard University Press, 1998.

Kleinschmidt, Harald. 'Using the Gun: Manual Drill and the Proliferation of Portable Firearms'. *Journal of Military History* 63, no. 3 (1999): 601–29.

Klötzer, Ralf. 'The Melchiorites and Münster'. In *A Companion to Anabaptism and Spiritualism, 1521–1700*, edited by John D. Roth and James M. Stayer, 217–56. Brill, 2007.

Kluge, Mathias, and Florian Dörschel. 'E fuero le prime si vedessero in questi paiesi'. *Vierteljahrschrift für Sozial- und Wirtschaftsgeschichte* 109 (2022): 479–95.

Knecht, R. J. *The French Civil Wars, 1562–1598*. Routledge, 2000.

Knecht, R. J. *Renaissance Warrior and Patron: The Reign of Francis I*. Revised edition. Cambridge University Press, 1996.

Knevel, Paul. *Burgers in het geweer: De schutterijen in Holland, 1550–1700*. Uitgeverij Verloren, 1994.

Krenn, Peter, Paul Kalaus and Bert Hall. 'Material Culture and Military History: Test-Firing Early Modern Small Arms'. *Material Culture Review* 42, no. 1 (1995): 101–9.

La Rocca, Donald J. 'Afonso "the African" and His Army: The Pastrana Tapestries as a Visual Encyclopedia for the Study of Arms and Armor'. In *The Invention of*

Glory: Afonso V and the Pastrana Tapestries, 29–41. Fundación Carlos de Amberes / Ediciones El Viso, 2011.

LaTour, Bruno. *Pandora's Hope: Essays on the Reality of Science Studies*. Harvard University Press, 1999.

Le armi degli Estensi: la collezione di Konopiště. Fabbri Editori, 1986.

Le Gall, Jean-Marie. *L'honneur perdu de François Ier: Pavie 1525*. Payot, 2015.

Loades, David. 'Early Reception'. *The Unabridged Acts and Monuments Online*. Humanities Research Institute, University of Sheffield, 2011. https://www.dhi.ac.uk/foxe/index.php?realm=more&gototype=&type=essay&book=essay7.

Lundén, Staffan. 'Distorting History in the Restitution Debate. Dan Hicks's *The Brutish Museums* and Fact and Fiction in Benin Historiography'. *International Journal of Cultural Property* 31 (2024): 202–25.

Lynn, John A. 'The *trace italienne* and the Growth of Armies'. In *The Military Revolution Debate: Readings on the Military Transformation of Early Modern Europe*, edited by Clifford J. Rogers, 169–200. Westview Press, 1995.

Madden, Amanda. *Civil Blood: Vendetta Violence and the Civic Elites in Early Modern Italy*. Cornell University Press, 2025.

Magli, Elena. 'Walking the Streets and Gathering in Squares: Civic Guarding and Defending in Late Medieval Basel'. In *Cities in Arms: Urban Martial Culture in Late Medieval and Early Modern Europe*, edited by Regula Schmid and Daniel Jaquet. Boydell and Brewer, forthcoming.

Mallett, Michael. 'Condottieri and Captains in Renaissance Italy'. In *The Chivalric Ethos and the Development of Military Professionalism*, edited by D. J. B. Trim, 67–88. Brill, 2003.

Mallett, Michael. *Mercenaries and Their Masters: Warfare in Renaissance Italy*. Bodley Head, 1974.

Mallett, Michael, and J. R. Hale. *The Military Organization of a Renaissance State: Venice c. 1400 to 1617*. Cambridge University Press, 1984.

Mallett, Michael, and Christine Shaw. *The Italian Wars 1494–1559: War, State and Society in Early Modern Europe*. Pearson, 2012.

Mancini, Matteo. *Tiziano e le corti Asburgo nei documenti degli archivi spagnoli*. Istituto veneto di scienze, lettere ed arti, 1998.

Mann, James G. 'The Lost Armoury of the Gonzagas. Part I'. *Archaeological Journal* 95 (1938): 239–336.

Mann, James G. 'The Lost Armoury of the Gonzagas. Part II: The Libro Aquila'. *Archaeological Journal* 100 (1948): 16–127.

Marchesi, Giampetro. 'Le miniere di ferro della Val Trompia'. In *Heavy Metal: Acciaio, oro e polvere da sparo nel Museo delle Armi "Luigi Marzoli" di Brescia*, edited by Marco Merlo, 25–35. Skira, 2022.

Markey, Lia. *Imagining the Americas in Medici Florence*. Pennsylvania State University Press, 2016.

Marshall, Peter. 'Robert Pakington'. In *Oxford Dictionary of National Biography*. Oxford University Press, 2004.

Martoccio, Michael. '"The Place for Such Business": The Business of War in the City of Genoa, 1701–1714'. *War in History* 29 (2021): 302–22.

Mayer, Thomas. 'Reginald Pole'. In *Oxford Dictionary of National Biography*. Oxford University Press, 2004.

Mayer, Thomas. *Reginald Pole: Prince and Prophet*. Cambridge University Press, 2000.

McCray, Patrick W. *Glassmaking in Renaissance Venice: The Fragile Craft*. Ashgate, 1999.

McNeill, William H. *Keeping Together in Time: Dance and Drill in Human History*. Harvard University Press, 1995.

Merlo, Marco, ed. *Heavy Metal: Acciaio, oro e polvere da sparo nel Museo delle Armi "Luigi Marzoli" di Brescia*. Skira, 2022.

Mocarelli, Luca, and Giulio Ongaro. 'Weapons Production in the Republic of Venice in the Early Modern Period: The Manufacturing Centre of Brescia between Military Needs and Economic Equilibrium'. *Scandinavian Economic History Review* 65 (2017): 231–42.

Moiloa, Molemo. *Reclaiming Restitution: Centering and Contextualizing the African Narrative*. Africa No Filter, 2022. https://openrestitution.africa/wp-content/uploads/2022/09/ANF-Report-Main-Report.pdf.

Moitzi, Andreas. 'Vienna's Military Organization'. In *Cities in Arms: Urban Martial Culture in Late Medieval and Early Modern Europe*, edited by Regula Schmid and Daniel Jaquet. Boydell and Brewer, forthcoming.

Montanari, Daniele. *Quelle terre di là dal Mincio: Brescia e il contado in età veneta*. Grafo, 2005.

Morin, Marco. *Armi antiche*. Mondadori, 1982.

Morin, Marco. 'La produzione delle armi da fuoco a Gardone V. T.'. In *Armi e cultura nel bresciano, 1420–1870*. Ateneo di Brescia, 1981.

Morin, Marco. 'The Origins of the Wheellock: A German Hypothesis. An Alternative to the Italian Hypothesis'. In *Art, Arms and Armour: An International Anthology*, vol. 1, *1979–80*, edited by Robert Held, 80–99. Acquafresca Editore, 1979.

Morin, Marco, and Robert Held. *Beretta: The World's Oldest Industrial Dynasty*. Aquafresca, 1980.

Muccini, Ugo. *The Salone del Cinquecento of Palazzo Vecchio*. Le Lettere, 1990.

Mulcahy, Rosemarie. 'Enea Vico's Proposed Triumphs of Charles V'. *Print Quarterly* 19 (2002): 331–40.

Museo delle Armi e della Tradizione Armeria di Gardone Valtrompia. 'Lavorazione Artigianale'. Archived by Wayback Machine on 25 May 2022. https://web.archive.org/web/20220525025223/http://www.museodellearmi.net/index.php?section=artigianale&m=percorso&pan=1.

Najemy, John M. 'Machiavelli and Cesare Borgia: A Reconsideration of Chapter 7 of *The Prince*'. *Review of Politics* 75, no. 4 (2013): 539–56.

Nayar, Sheila J. *Renaissance Responses to Technological Change*. Palgrave Macmillan, 2019.

Ndiaye, Noémie, and Lia Markey, eds. *Seeing Race before Race: Visual Culture and the Racial Matrix in the Premodern World*. ACMRS Press, 2023.

Nethersole, Scott. '*Armeggerie*, Wedding Chests and Battles in Fifteenth-century Florence'. *Renaissance Studies* 32 (2017): 282–304.

Neuschel, Kristen B. *Living by the Sword: Weapons and Material Culture in France and Britain, 600–1600*. Cornell University Press, 2020.

Nowak, Matthias, and André Gsell. *Handmade and Deadly: Craft Production of Small Arms in Nigeria*. Small Arms Survey, 2018.

Nuti, Giovanni. 'Adamo Centurione'. *Dizionario biografico degli italiani*. Treccani, 1979. https://www.treccani.it/enciclopedia/adamo-centurione_(Dizionario-Biografico).

Odorici, Federico, *Cenni storici sulle fabbriche d'armi della provincia bresciana*. [Publisher unknown], 1860.

Oestreich, Gerhard. *Neostoicism and the Early Modern State*. Edited by Brigitta Oestreich and H. G. Koenigsberger. Translated by David McLintock. Cambridge University Press, 1982.

Oleksiak, Wojciech. 'The Monk Who Stopped Bullets with Silk: Inventing the Bulletproof Vest'. *Culture.PL*, 4 January 2016. https://culture.pl/en/article/the-monk-who-stopped-bullets-with-silk-inventing-the-bulletproof-vest.

Panciera, Walter. 'Venetian Gunpowder in the Second Half of the Sixteenth Century: Production, Storage, Use'. In *Gunpowder, Explosives and the State: A Technological History*, edited by Brenda J. Buchanan, 93–122. Ashgate, 2006.

Panofsky, Erwin. *Problems in Titian: Mostly Iconographic*. New York University Press, 1969.

Paredes, Cecilia. 'The Confusion of the Battlefield: A New Perspective on the Tapestries of the Battle of Pavia (*c.* 1525–1531)'. *RIHA Journal* (2014), Article 01012.

Parker, Geoffrey. 'The "Military Revolution"—a Myth?' In *The Military Revolution Debate: Readings on the Military Transformation of Early Modern Europe*, edited by Clifford J. Rogers, 37–54. Westview Press, 1995.

Parrott, David. *The Business of War: Military Enterprise and Military Revolution in Early Modern Europe*. Cambridge University Press, 2012.

Pasero, Carlo. *La partecipazione bresciana alla Guerra di Cipro e alla Battaglia di Lepanto (1570–73)*. Geroldi, 1954.

PatER–Catalogo del Patrimonio culturale dell'Emilia-Romagna. 'Attentato a San Carlo'. Accessed 28 August 2025. https://bbcc.regione.emilia-romagna.it/pater/loadcard.do?id_card=50407&force=1.

Pedrocco, Filippo. *Titian: The Complete Paintings*. Thames and Hudson, 2001.

Pellegrini, Marco. *Le guerre d'Italia 1494–1530*. Il Mulino, 2009.

Pepper, Simon. 'Castles and Cannon in the Naples Campaign of 1494–95'. In *The French Descent into Renaissance Italy*, edited by David Abulafia. Ashgate, 1995.

Pepper, Simon. 'The Face of the Siege: Fortification, Tactics and Strategy in the Early Italian Wars'. In *Italy and the European Powers: The Impact of War, 1500–1530*, edited by Christine Shaw, 33–56. Brill, 2006.

Pepper, Simon. 'Patriots and Partisans: Popular Resistance to the Occupation of the Venetian *Terraferma* by the Forces of the League of Cambrai'. In *Warfare and Politics: Cities and Government in Renaissance Tuscany and Venice*, edited by H. C. Butters and Gabriele Neher, 79–103. Amsterdam University Press, 2020.

Pepper, Simon, and Nicholas Adams. *Firearms and Fortifications: Military Architecture and Siege Warfare in Sixteenth-century Siena*. Chicago University Press, 1986.

Pessina, Jacopo. *L'organizzazione militare della repubblica di Siena, 1524–1555*. Pisa University Press, 2022.

Peterson, Jeanette Favrot, and Kevin Terraciano, eds. *The Florentine Codex: An Encyclopedia of the Nahua World in Sixteenth-century Mexico*. University of Texas Press, 2019.

Pezzolo, Luciano. 'La "rivoluzione militare": una prospettiva italiana 1400–1700'. In *Militari in età moderna: la centralità di un tema di confine; Milano, 20 giugno 2004*, edited by Alessandra Dattero and Stefano Levati, 15–62. Cisalpino, 2006.

Pezzolo, Luciano. 'Republics and Principalities in Italy'. In *The Rise of Fiscal States: A Global History, 1500–1914*, edited by Bartolomé Yun-Casalilla and Patrick K. O'Brien with Francisco Comín Comín, 267–84. Cambridge University Press, 2012.

Phillips, Barnaby. *Loot: Britain and the Benin Bronzes*. Oneworld, 2021.

Phillips, Gervase. 'Longbow and Hackbutt: Weapons Technology and Technology Transfer in Early Modern England'. *Technology and Culture* 40 (1999): 576–93.

Piscini, Angela. 'Degli Agostini, Niccolò'. In *Dizionario Biografico degli Italiani*. Treccani, 1988. https://www.treccani.it/enciclopedia/niccolo-degli-agostini_(Dizionario-Biografico)/.

Pizzamiglio, Pierluigi, ed. *Atti della giornata di studio in memoria di Niccolò Tartaglia nel 450° anniversario della sua morte, 13 dicembre 1557–2007*. Ateneo di Brescia, 2007.

Quarenghi, Cesare. *Tecno-cronografia delle armi da fuoco italiane*. 2 vols. Nobile, 1880–81.

Racinaire. 'File:Scène de bataille sur le tombeau de François Ier dans la basilique Saint-Denis (2).JPG'. Wikipedia Commons. Last modified 29 August 2012. https://commons.wikimedia.org/wiki/File:Sc%C3%A8ne_de_bataille_sur_le_tombeau_de_Fran%C3%A7ois_Ier_dans_la_basilique_Saint-Denis_(2).JPG.

Radburn, Nicholas. 'The British Gunpowder Industry and the Transatlantic Slave Trade'. *Business History Review* 97 (2023): 363–84.

Raggio, Osvaldo. *Feuds and State Formation, 1550–1700: The Backcountry of the Republic of Genoa*. Palgrave Macmillan, 2018.

Rao, Ida Giovanna. 'On the Reception of the Florentine Codex: The First Italian Translation'. In *The Florentine Codex: An Encyclopedia of the Nahua World in Sixteenth-century Mexico*, edited by Jeanette Favrot Peterson and Kevin Terraciano, 37–44. University of Texas Press, 2019.

Raudzens, George, ed. *Technology, Disease, and Colonial Conquests, Sixteenth to Eighteenth Centuries: Essays Reappraising the Guns and Germs Theories*. Brill, 2001.

Rebecchini, Guido. *"Un altro Lorenzo": Ippolito de' Medici tra Firenze e Roma (1511–1535)*. Marsilio Editori, 2010.

Riccadonna, Alberto. 'Dal minerale di ferro all'arcobugio. L'arte di fare canne a Gardone'. In *Heavy Metal: Acciaio, oro e polvere da sparo nel Museo delle Armi "Luigi Marzoli" di Brescia*, edited by Marco Merlo, 57–69. Skira, 2022.

Richardson, Brian. *Printing, Writers and Readers in Renaissance Italy*. Cambridge University Press, 1999.

Richardson, Catherine, Tara Hamling and David Gaimster. *The Routledge Handbook of Material Culture in Early Modern Europe*. Routledge, 2017.

Ridley, Nick. *William the Silent and the Dutch Revolt: Comparative Starting Points and Triggering of Insurgencies*. Routledge, 2022.

Roberts, Michael. 'The Military Revolution, 1560–1660'. In *The Military Revolution Debate: Readings on the Military Transformation of Early Modern Europe*, edited by Clifford J. Rogers, 13–36. Westview Press, 1995.

Roberts-Jones, Philippe, and Françoise Roberts-Jones. *Bruegel*. Flammarion, 2012.

Robertson, Clare. *"Il Gran Cardinale": Alessandro Farnese, Patron of the Arts*. Yale University Press, 1992.

Roelofson, Mathijs. 'Fighting for the Town: Military Duties in Fifteenth-century Fribourg en Nuithonie from the Citizens' Perspective'. In *Cities in Arms: Urban Martial Culture in Late Medieval and Early Modern Europe*, edited by Regula Schmid and Daniel Jaquet. Boydell and Brewer, forthcoming.

Rogers, Clifford J., ed. *The Military Revolution Debate: Readings on the Military Transformation of Early Modern Europe*. Westview Press, 1995.

Romani, Marzio A. *Le corti farnesiane di Parma e Piacenza (1545–1622)*. Vol. 1, *Potere e società nello stato farnesiano*. Bulzoni, 1978.

Romanoni, Fabio, and Fabio Bargigia. 'La diffusione delle armi da fuoco nel dominio visconteo (secolo XIV)'. *Revista Universitaria de Historia Militar* 6 (2017): 136–55.

Rose, Colin. *A Renaissance of Violence: Homicide in Early Modern Italy*. Cambridge University Press, 2019.

Rota, Martin. 'Il Vero ritratto di Zarra et di Sebenico', 1570. Newberry Digital Collections. Accessed 28 August 2025. https://collections.newberry.org/asset-management/2KXJ8ZSAPPII1.

Royal Armouries. 'Matchlock Military Musket–Gardone Type–about 1540'. Accessed 28 August 2025. https://royalarmouries.org/collection/object/object-31168.

Ruff, Julius R. *Violence in Early Modern Europe*. Cambridge University Press, 2001.

Sabatti, Carlo. 'La Valtrompia nel '500'. In *Valtrompia nella storia*, edited by F. Bossini and M. Galeri, 157–213. Compagnia della Stampa, 2007.

Sani, Elisa Paola. 'War on a Plate: The Battle of Mühlberg on a Maiolica Dish at the Wallace Collection, London'. In *Pots, Prints and Politics: Ceramics with an Agenda, from the 14th to the 20th Century*, edited by Patricia F. Ferguson, 42–50. British Museum Research Publication 229, 2021.

Satia, Priya. *Empire of Guns: The Violent Making of the Industrial Revolution*. Duckworth, 2018.

Saunders, A. C. de C. M. *A Social History of Black Slaves and Freedmen in Portugal, 1441–1555*. Cambridge University Press, 1982.

Scalini, Mario. 'The Weapons of Lorenzo de' Medici: An Examination of the Inventory of the Medici Palace in Florence Drawn Up upon the Death of Lorenzo the Magnificent in 1492'. In *Art, Arms and Armour: An International Anthology*, vol. 1, *1979–80*, edited by Robert Held, 12–29. Acquafresca Editore, 1979.

Schalkhausser, Erwin. 'Peter Peck, the Emperor's Gunsmith'. In *Art, Arms and Armour: An International Anthology*, vol. 1, *1979–80*, edited by Robert Held, 183–203. Acquafresca Editore, 1979.

Schilling, Heinz. '"History of Crime" or "History of Sin"? Some Reflections on the Social History of Early Modern Church Discipline'. In *Politics and Society in Reformation*

Europe: Essays for Sir Geoffrey Elton on his Sixty-Fifth Birthday, edited by E. I. Kouri and Tom Scott, 289–310. Macmillan, 1987.

Schmid, Regula, and Daniel Jaquet, eds. *Cities in Arms: Urban Martial Culture in Late Medieval and Early Modern Europe*. Boydell and Brewer, forthcoming.

Schmitz-von Ledebur, Katja. 'Emperor Charles V Captures Tunis: A Unique Set of Tapestry Cartoons'. *Studia Bruxellae* 11 (2019): 387–404.

Schurink, Fred. 'War, What Is It Good For? Sixteenth-century English Translations of Ancient Texts on Warfare'. In *Renaissance Cultural Crossroads: Translation, Print and Culture in Britain, 1473–1640*, edited by Sara Barker and Brenda Hosington, 121–38. Brill, 2013.

Schwoerer, Lois G. *Gun Culture in Early Modern England*. University of Virginia Press, 2016.

Serck, Luc. 'Entourage de Henri Bles, Paysage avec les travaux de la mine'. In *Autour de Henri Bles*, edited by Jacques Toussaint. Société archéologique de Namur, 2000.

Shaw, Christine. 'Concepts of *Libertà* in Renaissance Genoa'. In *Communes and Despots in Medieval and Renaissance Italy*, edited by John E. Law and Bernadette Paton, 177–92. Routledge, 2016.

Shaw, Christine. *The Political Role of the Orsini Family from Sixtus IV to Clement VII: Barons and Factions in the Papal States*. Nella Sede Dell'Istituto Palazzo Borromini, 2007.

Shaw, Christine. *The Politics of Exile in Renaissance Italy*. Cambridge University Press, 2000.

Shaw, Christine, ed. *Italy and the European Powers: The Impact of War, 1500–1530*. Brill, 2006.

Shaw, James E. 'The Informal Economy of Credit in Early Modern Venice'. *The Historical Journal* 61 (2018): 623–42.

Sherer, Idan. *The Scramble for Italy: Continuity and Change in the Italian Wars, 1494–1559*. Routledge, 2021.

Sherer, Idan. *Warriors for a Living: The Experience of the Spanish Infantry during the Italian Wars, 1494–1559*. Brill, 2017.

Sicca, Cinzia M. 'Consumption and Trade of Art between Italy and England in the First Half of the Sixteenth Century: The London House of the Bardi and Cavalcanti Company'. *Renaissance Studies* 16 (2002): 163–201.

Sicca, Cinzia M., and Louis Alexander Waldman, eds. *The Anglo-Florentine Renaissance: Art for the Early Tudors*. Yale Center for British Art; Paul Mellon Centre for Studies in British Art, 2012.

Silver, Larry. *Marketing Maximilian: The Visual Ideology of a Holy Roman Emperor*. Princeton University Press, 2008.

Silverman, David J. *Thundersticks: Firearms and the Violent Transformation of Native America*. Harvard University Press, 2016.

Small, M. 'Transformer and Influencer: Giovanni Battista Ramusio's Impact on Western European Geography'. *Journal of Early Modern Studies* 12 (2023): 39–54.

Small Arms Survey. 'Global Firearms Holdings'. 2018. Accessed 9 February 2025. https://www.smallarmssurvey.org/database/global-firearms-holdings.

Smith, Robert. *Warfare and Diplomacy in Pre-colonial West Africa*. 2nd ed. Currey, 1989.

Società Storica Lombarda. 'Giovanni Battista Porcellaga + Cecilia Gavardo'. In *Enciclopedia delle Famiglie Lombarde*. Accessed 28 August 2025. https://genealogie.societastoricalombarda.it/family.php?famid=F17750&ged=ssl.

Soykut, Mustafa. *Italian Perceptions of the Ottomans: Conflict and Politics through Pontifical and Venetian Sources*. Peter Lang, 2011.

Springer, Carolyn. *Armour and Masculinity in the Italian Renaissance*. Toronto University Press, 2010.

Stagno, Laura, and Borja Franco Llopis, eds. *Lepanto and Beyond: Images of Religious Alterity from Genoa and the Christian Mediterranean*. Leuven University Press, 2021.

Steadman, Philip. *Renaissance Fun: The Machines behind the Scenes*. UCL Press, 2021.

Stiber, Linda S., Elmer Eusman and Sylvia Albro. 'The Triumphal Arch and the Large Triumphal Carriage of Maximilian I: Two Oversized, Multi-block, 16th-century Woodcuts from the Studio of Albrecht Durer'. *The Book and Paper Group Annual* 14 (1995). https://cool.culturalheritage.org/coolaic/sg/bpg/annual/v14/bp14-07.html.

Stracke, Richard, with Claire Stracke. 'The Palazzo Bellomo Altarpiece of St. Barbara'. Christian Iconography. Accessed 25 October 2024. https://www.christianiconography.info/sicily/barbaraBellomo.portrait.html.

Strauss, Walter L., ed. *The Illustrated Bartsch, 10 (Commentary)*. Abaris Books, 1978.

Strickland, Matthew, and Robert Hardy. *The Great Warbow: From Hastings to the Mary Rose*. Haynes, 2011.

Strunck, Christina. 'The Barbarous and Noble Enemy: Pictorial Representations of the Battle of Lepanto'. In *The Turk and Islam in the Western Eye, 1450–1750: Visual Imagery before Orientalism*, edited by James G. Harper, 217–40. Ashgate, 2011.

Suckale-Redlefsen, Gude. *Mauritius, der heilige Mohr / The Black Saint Maurice*. Menil Foundation, 1987.

Sutherland, N. M. 'The Assassination of François Duc de Guise, February 1563'. *Historical Journal* 24 (1981): 279–95.

Sutherland, N. M. *Princes, Politics and Religion*. Hambledon Press, 1984.

Tagliabue, John. 'Small Steel Mills Feel Pinch in Italy'. *New York Times*, 2 January 1982. https://www.nytimes.com/1982/01/02/business/small-steel-mills-feel-pinch-in-italy.html.

Tanis, James, and Daniel Horst. *Images of Discord: A Graphic Interpretation of the Opening Decades of the Eighty Years War*. W. B. Eerdmans for Bryn Mawr College Library, 1993.

Targioni-Tozzetti, Giovanni. *Notizie sulla storia delle scienze fisiche in Toscana*. Biblioteca Palatina, 1852.

Taylor, F. L. *The Art of War in Italy, 1494–1529*. Cambridge University Press, 1921.

Terraciano, Kevin. 'Reading between the Lines of Book 12'. In *The Florentine Codex: An Encyclopedia of the Nahua World in Sixteenth-century Mexico*, edited by Jeanette Favrot Peterson and Kevin Terraciano, 45–62. University of Texas Press, 2019.

Thommen, Lukas. *An Environmental History of Ancient Greece and Rome*. Cambridge University Press, 2012.

Thompson, I. A. A. '"Money, Money, and Yet More Money!" Finance, the Fiscal-State, and the Military Revolution. Spain, 1500–1650'. In *The Military Revolution Debate: Readings on the Military Transformation of Early Modern Europe*, edited by Clifford J. Rogers, 273–98. Westview Press, 1995.

Thornton, Dora. *The Scholar in His Study*. Yale University Press, 1997.

Thornton, Peter. *The Italian Renaissance Interior*. Weidenfeld and Nicolson, 1991.

Tlusty, B. Ann. *The Martial Ethic in Early Modern Germany: Civic Duty and the Right of Arms*. Palgrave Macmillan, 2011.

Townsend, Camilla. *Fifth Sun: A New History of the Aztecs*. Oxford University Press, 2019.

Tryner, Joe. '"The Rascal with His Firestick": Gun Culture and Firearms Violence in Sixteenth-century Bologna'. Unpublished PhD thesis, University of Sheffield, 2024.

United Nations Office for Disarmament Affairs. 'Arms Trade'. Accessed 1 July 2022. https://www.un.org/disarmament/convarms/att.

Valseriati, Enrico. *Tra Venezia e l'Impero: dissenso e conflitto politico a Brescia nell'età di Carlo V*. FrancoAngeli, 2016.

Van Nimwegen, Olaf. 'The Transformation of Army Organisation in Early-modern Western Europe, c. 1500–1789'. In *European Warfare, 1350–1750*, edited by Frank Tallett and D. J. B. Trim, 159–80. Cambridge University Press, 2012.

Vannucci, Alessandra Baroni. *Jan Van Der Straet detto Giovanni Stradano: flandrus pictor et inventor*. Jandi Sapi Editore, 1997.

Wethey, Harold E. *The Paintings of Titian II: The Portraits*. Phaidon, 1971.

Williams, Ann. 'Mediterranean Conflict'. In *Süleyman the Magnificent and His Age: The Ottoman Empire in the Early Modern World*, edited by Metin Kunt and Christine Woodhead, 39–54. Routledge, 1995.

Wilson, Peter H., and Marianne Klerk. 'The Business of War Untangled: Cities as Fiscal-Military Hubs in Europe (1530s–1860s)'. *War in History* 29 (2022): 80–103.

Wilson, Robert L. *The World of Beretta*. Skyhorse, 2015.

Wood, James B. *The King's Army: Warfare, Soldiers and Society during the Wars of Religion in France, 1562–76*. Cambridge University Press, 1996.

Woolfson, Jonathan. *Padua and the Tudors: English Students in Italy, 1485–1603*. University of Toronto Press, 1998.

Zan, Luca, ed. *The Venice Arsenal: Between History, Heritage and Re-use*. Routledge, 2022.

Zanelli, Agostino. 'Un contratto di acquisto di archibugi bresciani per conto della Santa Sede nel 1571'. *Brescia nelle industry e nei commerci* 5, no. 3 (March 1925): 299–302.

INDEX

Page numbers in italic indicate illustrations.

A NOTE ON THE TYPE

THIS BOOK has been composed in Miller, a Scotch Roman typeface designed by Matthew Carter and first released by Font Bureau in 1997. It resembles Monticello, the typeface developed for The Papers of Thomas Jefferson in the 1940s by C. H. Griffith and P. J. Conkwright and reinterpreted in digital form by Carter in 2003.

Pleasant Jefferson ("P. J.") Conkwright (1905–1986) was Typographer at Princeton University Press from 1939 to 1970. He was an acclaimed book designer and AIGA Medalist.

The ornament used throughout this book was designed by Pierre Simon Fournier (1712–1768) and was a favorite of Conkwright's, used in his design of the *Princeton University Library Chronicle.*